T0091737

METHODS IN MOLECULAR BIOLOGY™

Series Editor
John M. Walker
School of Life Sciences
University of Hertfordshire
Hatfield, Hertfordshire, AL10 9AB, UK

For other titles published in this series, go to
www.springer.com/series/7651

RNA and DNA Editing

Methods and Protocols

Edited by

Ruslan Aphasizhev

Department of Microbiology and Molecular Genetics,
School of Medicine, University of California, Irvine, CA, USA

☀ Humana Press

Editor
Ruslan Aphasizhev, Ph.D.
Department of Microbiology and Molecular Genetics
School of Medicine
University of California
Irvine, CA 92697
USA
ruslan@uci.edu

ISSN 1064-3745   e-ISSN 1940-6029
ISBN 978-1-61779-017-1   e-ISBN 978-1-61779-018-8
DOI 10.1007/978-1-61779-018-8
Springer New York Heidelberg London Dordrecht

Library of Congress Control Number: 2011921338

Printed on acid-free paper

Humana Press is part of Springer Science+Business Media (www.springer.com)

Foreword

RNA editing is a disparate field held together by history and friendships between the researchers. The term was coined by Robb Benne in his 1986 Cell paper in which he described the presence of four nonencoded uridylyl nucleotides in the middle of the mRNA for cytochrome oxidase subunit II in the trypanosomatid protist, *Crithidia fasciculata*. The striking thing was that these precise U-insertions at the RNA level precisely compensated for or "edited" an encoded −1 frameshift in the mitochondrial maxicircle DNA encoded gene. In 1989, Janet Shaw and I tried to define RNA editing as "any modification of the sequence of an mRNA molecule within coding regions, except for splicing of introns." This definition encompassed a single C to U substitution in the mRNA for mammalian apolipoprotein B (apoB), specific A to I modifications in the mRNA for the glutamate receptor, multiple C to U substitutions in plant chloroplast and mitochondrial mRNAs, and even insertions of G residues within coding regions of negative strand RNA viruses. But it failed to cover specific nucleotide changes in tRNAs from *Acanthamoeba* and marsupials, and the most diverse and striking phenomenon of all, specific insertions of multiple C and dinucleotide residues and other modifications in all mitochondrial transcripts in *Physarum*, including rRNAs. The sexy term, RNA editing, was hijacked (just joking) to describe all of these disparate genetic events, in spite of the growing realization that quite different mechanisms were involved.

A book on RNA editing edited by Benne was published in 1994. The actual birth of the RNA editing research field can probably be traced to a meeting in Albany organized by Harold Smith the same year, which led to the establishment of a Gordon Conference on RNA editing in 1989 and thereby made the field respectable. This field was the "flavor of the week" for a while, but has had its ups and downs as is the case for any field of research. Even as the realization grew that the various editing phenomena were only related by the name, the field held together due to its scientific comaraderie and subsequent cross fertilization of ideas which, I believe, has led to many really exciting and unexpected discoveries, such as for example cytidine deaminase-induced DNA editing which is involved in the generation of antibody diversity. The field experienced some intense soul searching when it debated encompassing the large well-established field of nucleotide modifications in general as a type of "editing." The decision to do this not only provided the editing field with some very smart colleagues but also proved highly beneficial to the mechanistic understanding of all types of editing.

In 1990, Beat Blum, Norbert Bakalara, and I uncovered the mechanism of the U-insertion/deletion editing in trypanosome mitochondria by discovering a novel class of small mitochondrial RNAs which we termed "guide RNAs" since they guided the editing machinery to specific sites in the mRNAs. The solution to the precise site specificity of this type of editing was simply base-pairing in *trans*. The site specificity of A to I editing in neurons also proved to be due to base-pairing, except the complementary sequence is *in cis* downstream of the editing site. This same base-pairing mechanism turned out to be used by the small siRNAs which mediate the cleavage of RNAs in the Nobel Prize winning RNA interference phenomenon, which was discovered in 1998, and also by the hitherto mysterious yeast snoRNAs that determine the sites of pseudoU formation and

$2'$-O-methylation in rRNAs. Alas, the latter two fields never directly acknowledged their debt to the editing paradigm, but did so inadvertently by using the term, guide RNAs, to describe the siRNAs and snoRNAs. But base-pairing site specificity was not the answer to other types of editing and nucleotide modifications. The precise apoB mRNA C to U change was due to the recognition of an upstream "mooring sequence" by a deamination protein together with other factors, and the specificity of C to U changes in plant organel-lar RNAs was also due to the recognition of specific sequences by proteins. The viral G addition phenomenon turned out to be due to "stuttering" of a polymerase. And the incredible *Physarum* C-insertion phenomenon is cotranscriptional also with some sequence specificity.

A clear indication of the state of advancement and general acceptance of any field of research is to have a "*Methods*" book published, which describes "cookbook" procedures to repeat some of the most interesting discoveries in your own laboratory. Hence, the publication of this *Methods in Molecular Biology* book (and a *Methods in Enzymology* book in 2007 edited by Jonatha Gott) is to be celebrated. There are 16 chapters describing state-of-the-art research in trypanosome U-insertion/deletion editing, A to I editing in *Drosophila*, C to U RNA and DNA editing, C to U editing in plant chloroplasts and mito-chondria and RNA modifications.

Larry Simpson

Preface

The term "RNA editing," coined in 1986 to describe the insertion of four uridines into a mitochondrial transcript in *Trypanosoma brucei*, has evolved into a collective definition of processes that change RNA nucleotide sequence. Setting these pathways apart from splicing, 5′ capping, or 3′ extensions is uncomplicated while drawing a clear distinction from RNA modifications is less so. Spread throughout the Eukarya, editing creates genetic information de novo, alters decoding capacity, influences structure, stability, export, and other aspects of nucleic acids metabolism by inserting, deleting, adding, or modifying nucleotides. Evolutionarily unrelated, although sometimes confined to the same organism or organelle, editing events occur cotranscriptionally or posttranscriptionally. Mechanistically, RNA editing reactions include guide RNA-directed cascades of nucleolytic and phosphoryl transfer reactions, RNA polymerase stuttering, site-specific deamination, 3′–5′ polymerization, and others. A more recent but most exciting development, DNA editing is emerging as the key component of antibody gene diversification and antiviral defense.

Such diversity of organisms and editing types stimulated development of many unique genetic, molecular, biochemical, and computational approaches. The purpose of this volume is to introduce methods developed over the last few years to study the diversity of editing substrates, mechanisms of specificity, and functions of RNA and DNA editing enzymes and complexes.

I wish to express my sincere gratitude to the authors for their contributions and continued enthusiasm for our filed. This volume is dedicated to Rob Benne in appreciation of his seminal discovery.

Ruslan Aphasizhev

Contents

Contributors

JUAN D. ALFONZO • *Department of Microbiology, The Ohio State University, Columbus OH, USA*

RUSLAN APHASIZHEV • *Department of Microbiology & Molecular Genetics, School of Medicine, University of California, Irvine, CA, USA*

VALERIE BLANC • *Department of Medicine, Washington University School of Medicine, St. Louis, MO, USA*

RALPH BOCK • *Max Planck Institute of Molecular Plant Physiology, Potsdam-Golm, Germany*

CORDULA BÖHM • *Department of Genetics, Darmstadt University of Technology, Darmstadt, Germany*

YA-LIN CHIU • *Department of Medicine, Gladstone Institute of Virology and Immunology, University of California, San Francisco, CA, USA*

NICHOLAS O. DAVIDSON • *Department of Medicine, Washington University School of Medicine, St. Louis, MO, USA*

SCOTT DEWELL • *Genomics Resource Center, The Rockefeller University, New York, NY, USA*

KIRK W. GASTON • *Department of Microbiology, The Ohio State Center for RNA Biology, The Ohio State University, Columbus, OH, USA*

SELENA GELL • *Molecular Biology Cellular Biology and Biochemistry, Brown University, Providence, RI, USA*

MONIKA M. GOLAS • *The Water and Salt Research Center, Institute of Anatomy, Aarhus University, Århus C, Denmark*

H. ULRICH GÖRINGER • *Department of Genetics, Darmstadt University of Technology, Darmstadt, Germany*

MARK HELM • *Department of Chemistry, Institute of Pharmacy and Molecular Biotechnology, Ruprecht-Karls Universität Heidelberg, Heidelberg, Germany; Institute of Pharmacy and Biochemistry, Johannes Gutenberg-University Mainz, Mainz, Germany*

MARTIN HENGESBACH • *Institute of Pharmacy and Molecular Biotechnology, University of Heidelberg, Heidelberg, Germany*

CHAO HUANG • *Department of Biochemistry and Biophysics, University of Rochester Medical Center, Rochester, NY, USA*

JAMES E.C. JEPSON • *Molecular Biology Cellular Biology and Biochemistry, Brown University, Providence, RI, USA*

JOHN KARIJOLICH • *Department of Biochemistry and Biophysics, University of Rochester Medical Center, Rochester, NY, USA*

LIAM P. KEEGAN • *MRC Human Genetics Unit, Institute of Genetics and Molecular Medicine, Western General Hospital, Edinburgh, UK*

RICHARD H. LATHROP • *Department of Computer Science, School of Information and Computer Sciences, Institute for Genomics and Bioinformatics, University of California, Irvine, CA, USA*

VALERIE MANN • *MRC Centre for Regenerative Medicine, University of Edinburgh, Edinburgh, UK*

MADELEINE MEUSBURGER • *Institute of Pharmacy and Molecular Biotechnology, University of Heidelberg, Heidelberg, Germany*

SHULAMIT MICHAELI • *The Mina and Everard Goodman Faculty of Life Sciences and the Advanced Materials and Nanotechnology Institute, Bar-Ilan University, Ramat-Gan, Israel*

MICHAEL S. NEUBERGER • *Medical Research Council Laboratory of Molecular Biology, Cambridge, UK*

BRENDON NOBLE • *MRC Centre for Regenerative Medicine, University of Edinburgh, Edinburgh, UK*

MARY A. O'CONNELL • *MRC Human Genetics Unit, Institute of Genetics and Molecular Medicine, Western General Hospital, Edinburgh, UK*

F. NINA PAPAVASILIOU • *Laboratory of Lymphocyte Biology, The Rockefeller University, New York, NY, USA*

CRISTINA RADA • *Medical Research Council Laboratory of Molecular Biology, Cambridge, UK*

ARUNA RAJA • *MRC Human Genetics Unit, Institute of Genetics and Molecular Medicine, Western General Hospital, Edinburgh, UK*

ROBERT A. REENAN • *Molecular Biology Cellular Biology and Biochemistry, Brown University, Providence, RI, USA*

GENE-ERROL RINGPIS • *Department of Microbiology and Molecular Genetics, School of Medicine, University of California, Irvine, CA, USA*

BRAD R. ROSENBERG • *Laboratory of Lymphocyte Biology, The Rockefeller University, New York, NY, USA*

STEPHANIE RUF • *Max-Planck-Institut für Molekulare Pflanzenphysiologie, Potsdam-Golm, Germany*

MASAYUKI SAKURAI • *Department of Chemistry and Biotechnology, Graduate School of Engineering, University of Tokyo, Tokyo, Japan*

BJOERN SANDER • *Stereology and Electron Microscopy Research Laboratory, Aarhus University, Århus C, Denmark*

JESSICA L. SPEARS • *Department of Microbiology, The Ohio State Center for RNA Biology, The Ohio State University, Columbus, OH, USA*

CYNTHIA J. STABER • *Molecular Biology Cellular Biology and Biochemistry, Brown University, Providence, RI, USA*

HOLGER STARK • *Research Group of 3D Electron Cryomicroscopy, Max-Planck-Institute for Biophysical Chemistry, Göttingen, Germany; Göttingen Centre for Molecular Biology, University of Göttingen, Göttingen, Germany*

HUI SUN • *MRC Human Genetics Unit, Institute of Genetics and Molecular Medicine, Western General Hospital, Edinburgh, UK*

TSUTOMU SUZUKI • *Department of Chemistry and Biotechnology, Graduate School of Engineering, University of Tokyo, Bunkyo-ku, Tokyo, Japan*

MIZUKI TAKENAKA • *Molekulare Botanik, Universität Ulm, Ulm, Germany*

CHAIM WACHTEL • *The Mina and Everard Goodman Faculty of Life Sciences and the Advanced Materials and Nanotechnology Institute, Bar-Ilan University, Ramat-Gan, Israel*

MENG WANG • *Medical Research Council Laboratory of Molecular Biology, Cambridge, UK*

YI-TAO YU • *Department of Biochemistry and Biophysics, University of Rochester Medical Center, Rochester, NY, USA*

ANJA ZEHRMANN • *Molekulare Botanik, Universität Ulm, Ulm, Germany*

Part I

Uracil Insertion/Deletion RNA Editing in Mitochondrion of *Trypanosoma brucei*

Chapter 1

Three-Dimensional Reconstruction of *Trypanosoma brucei* Editosomes Using Single-Particle Electron Microscopy

H. Ulrich Göringer, Holger Stark, Cordula Böhm, Bjoern Sander, and Monika M. Golas

Abstract

RNA editing within the mitochondria of kinetoplastid protozoa is performed by a multicomponent macromolecular machine known as the editosome. Editosomes are high molecular mass protein assemblies that consist of about 15–25 individual polypeptides. They bind pre-edited transcripts and convert them into translation-competent mRNAs through a biochemical reaction cycle of enzyme-catalyzed steps. At steady-state conditions, several distinct complexes can be purified from mitochondrial detergent lysates. They likely represent RNA editing complexes at different assembly stages or at different functional stages of the processing reaction. Due to their low cellular abundance, single-particle electron microscopy (EM) represents the method of choice for their structural characterization. This chapter describes a set of techniques suitable for the purification and structural characterization of RNA editing complexes by single-particle EM. The RNA editing complexes are isolated from the endogenous pool of mitochondrial complexes by tandem-affinity purification (TAP). Since the TAP procedure results in the isolation of a mixture of different RNA editing complexes, the isolates are further subjected to an isokinetic ultracentrifugation step to separate the complexes based on their sedimentation behavior. The use of the "GraFix" protocol is presented that combines mild chemical cross-linking with ultracentrifugation. Different sample preparation protocols including negative staining, cryo-negative staining, and unstained cryotechniques as well as the single-particle image processing of electron microscopical images are described.

Key words: RNA editing, Editosome, *Trypanosoma brucei*, Tandem-affinity purification (TAP), GraFix, Surface plasmon resonance (SPR), Density gradient centrifugation, Electron microscopy (EM), Cryo-EM, Single-particle image processing

1. Introduction

Many cellular proteins act in complexes with other proteins or nucleic acids as part of high molecular mass macromolecular machines (1). These complexes typically assemble in multistep reaction pathways, and thus the composition and function of

Ruslan Aphasizhev (ed.), *RNA and DNA Editing: Methods and Protocols*, Methods in Molecular Biology, vol. 718,
DOI 10.1007/978-1-61779-018-8_1, © Springer Science+Business Media, LLC 2011

complexes largely depends on the assembly step. One such example is the RNA editing machinery in kinetoplastid protozoa such as African trypanosomes and *Leishmania*. The RNA editing complexes generate messenger ribonucleic acid (mRNA) molecules from immature pre-messenger RNA (pre-mRNA) by the insertion and/or deletion of exclusively uridylate residues (2, 3). In trypanosomes, more than 20 proteins are involved in the process and the catalytic complexes have been termed as editosomes. Different RNA editing complexes sedimenting between about 5 and 40 Svedberg units (S) have been identified.

For the three-dimensional (3D) structural analysis of such multicomponent complexes, single-particle transmission electron microscopy (EM) represents the method of choice (4, 5). In contrast to X-ray crystallography and nuclear magnetic resonance (NMR), single-particle EM can successfully deal with low sample concentrations (in case of the editosomes typically below 10 μg/ml) even in the case of transient and fragile assemblies (6). The basic principle of single-particle EM is the computational merging of many thousands of individual particles imaged in the electron microscope to reconstruct their 3D structure (7, 8). Due to their pronounced radiation sensitivity, biological samples have to be imaged at low-dose conditions. As a consequence, the raw EM images are noisy and signals need to be enhanced by computational averaging methods. In addition, the structural analysis of macromolecular assemblies is often hampered by the disintegration and/or aggregation of the complexes during purification and sample preparation, thereby limiting the applicability of single-particle EM on a number of macromolecular machines. However, a novel approach offers a promising protocol to tackle these limitations (9). Recent developments in single-particle EM allow the analysis of structural transitions and dynamics processes during the assembly and catalytic cycle of macromolecular machines on the 3D level.

2. Materials

2.1. TAP Purification of RNA Editing Complexes

1. SDM-79 medium (all media components from Invitrogen, Karlsruhe, Germany) supplemented with 10% (v/v) fetal calf serum (10).

2. pLEW100/TAP, a derivative of pLEW100 (11).

3. Primers used to amplify the coding region of TbMP42 (GenBank AF382335): TbMP42-5′-primer: CCGCTCGAGAT GAAGCGTGTTACTTCACATATTTCG; TbMP42-3′-primer: TGCTCTAGACACCCTCAACACTGACCCAAGCC.

4. Editing buffer: 20 mM HEPES-KOH, pH 7.5, 30 mM KCl, 10 mM Mg(OAc)$_2$, 0.5 mM DTT.

5. Lysis buffer: 20 mM HEPES-KOH pH 7.5, 30 mM KCl, 10 mM Mg(OAc)$_2$, 0.5 mM DTT, 0.6% (v/v) Nonidet-P40.

6. IPP150: 10 mM Tris–HCl, pH 8, 150 mM NaCl, 1% (v/v) Nonidet-P40.

7. IgG Sepharose (Amersham Biosciences, Freiburg, Germany) equilibrated in editing buffer.

8. TEV cleavage buffer: 10 mM Tris–HCl, pH 8, 150 mM NaCl, 1% (v/v) Nonidet-P40, 0.5 mM Na$_2$EDTA, 1 mM DTT.

9. TEV protease (BioPioneer Inc., San Diego, CA, USA).

10. Calmodulin binding buffer: 10 mM Tris–HCl, pH 8, 150 mM NaCl, 1% (v/v) Nonidet-P40, 10 mM 2-mercaptoethanol, 1 mM Mg(OAc)$_2$, 1 mM imidazol, 2 mM CaCl$_2$.

11. Calmodulin affinity resin (Stratagene, La Jolla, USA) equilibrated in editing buffer.

12. Calmodulin elution buffer: 10 mM Tris–HCl, pH 8, 150 mM NaCl, 1% (v/v) Nonidet-P40, 10 mM 2-mercaptoethanol, 1 mM Mg(OAc)$_2$, 1 mM imidazol, 5 mM EGTA.

13. RNasin©.

2.2. Biochemical Analysis

1. Calf intestine phosphatase.

2. Guanylyl transferase (Ambion Inc., Austin, TX, USA).

3. Polynucleotide kinase.

4. γ-[^{32}P]-ATP.

5. α-[^{32}P]-GTP.

6. Na [^{125}I] (Hartmann Analytic). See Note 1.

7. Chloramine-T (Acros Organics, Geel, Belgium) dissolved in 50 mM sodium phosphate, pH 7.5.

8. Na$_2$S$_2$O$_5$ (Acros Organics, Geel, Belgium) dissolved in 50 mM sodium phosphate, pH 7.5.

9. Beckman TLS55 rotor.

10. Low percentage buffer: 10% (v/v) glycerol, 20 mM HEPES-KOH pH 7.5, 30 mM KCl, 10 mM Mg(OAc)$_2$, 0.5 mM DTT.

11. High percentage buffer: 40% (v/v) glycerol, 20 mM HEPES-KOH, pH 7.5, 30 mM KCl, 10 mM Mg(OAc)$_2$, 0.5 mM DTT, 0.1% (v/v) glutaraldehyde.

12. Cushion buffer: 20 mM HEPES-KOH pH 7.5, 30 mM KCl, 10 mM Mg(OAc)$_2$, 0.5 mM DTT, 10% (v/v) glycerol.

2.3. Surface Plasmon Resonance

1. Oxidation buffer: 50 mM NaOAc, pH 4.8, 10 mM $MgCl_2$, 100 mM NaCl.

2. Coupling buffer: 100 mM sodium phosphate, pH 7.2, 150 mM NaCl, 50 mM $NaBH_3CN$.

3. IAsys surface plasmon resonance (SPR) instrument (NeoSensors, Sedgefield, UK).

4. Aminosilane-coated SPR reaction cuvette (NeoSensors, Sedgefield, UK).

2.4. GraFix Fractionation

1. Gradient preparation device, e.g., GradientMaster (Biocomp, Fredericton, NB, Canada).

2. Ultracentrifuge, rotors and centrifugation tubes: Sorvall or Beckman ultracentrifuge with suitable rotor with a capacity of about 13.2 ml per centrifugation tube (e.g., Sorvall TH-641).

3. Gradient fractionation device: custom-made needle/tube system connected to a peristaltic pump and a UV spectrophotometer (GE Healthcare, Buckinghamshire, UK) or tube piercing device (Brandel Isco tube piercer, Isco, Lincoln, USA).

4. EM-grade 25% (v/v) glutaraldehyde solution (Electron Microscopy Sciences, Hatfield, PA, USA). See Note 2.

2.5. Negative Staining and Cryo-negative Staining

1. Custom-made black plastic (polyoxymethylene) or Teflon block with holes of about 25–30 and 100–200 µl volume.

2. Home-made carbon film indirectly coated on mica. Carbon films can be prepared by evaporating carbon on freshly cleaved mica using a carbon evaporating device.

3. Copper EM grids covered with a perforated carbon film, either home-made or commercial, e.g., Quantifoil (Quantifoil Micro Tools, Jena, Germany) or C-flat (Protochips, Raleigh, NC, USA).

4. Liquid nitrogen storage container for cryogrids.

5. Negative staining solution, e.g., 2% (w/v) uranyl formate in water (see Note 3).

6. Glycerol-free buffer: 20 mM HEPES-KOH pH 7.5, 30 mM KCl, 10 mM $Mg(OAc)_2$, 0.5 mM DTT.

2.6. Unstained Cryopreparation

1. Freeze-plunging device (home-made or commercial, e.g., Vitrobot, FEI, Eindhoven, The Netherlands or Cryoplunge, Gatan, Pleasanton, CA, USA). See Note 4.

2. Glow-discharging device (home-made or Leica Microsystems, Wetzlar, Germany).

3. Buffer-exchange columns, e.g., Zeba spin columns (Pierce, Rockford, IL, USA).

4. Liquid N_2. Gaseous ethane. See Note 4.

5. Copper EM grids covered with a perforated carbon film, either home-made or commercial, e.g., Quantifoil (Quantifoil Micro Tools) or C-flat (Protochips).

6. Liquid nitrogen storage container for cryogrids.

2.7. Electron Microscopy and Single-Particle Image Processing

1. Electron cryomicroscope equipped with LaB_6 cathode or field emission gun (FEG), suitable lens, cryostage, and low-dose capabilities (e.g., FEI or JEOL, Tokyo, Japan).

2. For imaging at room temperature: Room temperature side-entry holder or loading system (e.g., FEI or JEOL).

3. For imaging under cryogenic conditions: Side-entry cryohol der (Gatan) or cryoloading system (FEI).

4. For imaging on a charge-coupled device (CCD) camera: Slow-scan CCD camera optimized for low-dose work (e.g., 4k × 4k CCD camera from Tietz, Gauting, Germany or Gatan, or FEI).

5. For imaging on photographic film: Kodak SO163 photographic film.

6. For imaging on photographic film: High-resolution scanner, e.g., Tango, Heidelberger Druckmaschinen, Heidelberg, Germany or CoolScan, Nikon, Tokyo, Japan.

7. Computer cluster for image processing and graphical work-station for particle selection, visualization of single-particle image processing intermediates, and 3D maps.

8. Image processing software, e.g., IMAGIC (http://www. imagescience.de), SPIDER (http://www.wadsworth.org/ spider_doc/spider/docs/spider.html), EMAN (http://blake. bcm.tmc.edu/eman/eman1), XMIPP (http://xmipp.cnb. csic.es), or FREALIGN (http://emlab.rose2.brandeis.edu/ grigorieff/downloads.html).

9. 3D visualization software, e.g., Amira (www.amiravis.com), Chimera (http://www.cgl.ucsf.edu/chimera/download. html), PyMOL (http://pymol.sourceforge.net).

3. Methods

3.1. TAP Purification of RNA Editing Complexes

Native RNA editing complexes were isolated by tandem-affinity purification (TAP) (12). For that we generated a trypanosome cell line that expresses a C-terminally TAP-tagged version of the mitochondrial protein TbMP42, which is an integral component of the *Trypanosoma brucei* editosome (13, 14). The open reading frame was amplified from genomic DNA of *T. brucei* strain 427 (15). The resulting PCR fragment was cloned into pLEW100/TAP, a derivative of pLEW100 (11). pLEW100/TAP contains the TAP-tag cassette of pBS1479 (12). The expression plasmid pLEW-TbMP42/TAP was transfected into *T. brucei* strain 29-13 to allow conditional expression of TbMP42/TAP by adding tetracyclin (tet) to the culture medium (11). Clonal cell lines of 29-13-MP42/TAP were established by limited dilution. Expression of the 64.3 kDa (577 amino acids) TbMP42/TAP protein was verified by Western blotting.

1. Grow 29-13 TbMP42/TAP trypanosomes in SDM-79 medium supplemented with 10% (v/v) FCS, 7.5 mg/l hemin, and 50 U/ml penicillin/streptomycin at 27°C.

2. At a cell density of $1–2 \times 10^7$ cells/ml induce expression of TbMP42/TAP by adding 1 μg/ml tetracycline to the culture medium. Incubate for 72–96 h.

3. Harvest parasites from 20 l of cell culture at cell densities between 1 and 2×10^7 cells/ml.

4. Isolate mitochondrial vesicles at isotonic conditions by N_2 cavitation (16).

5. Lyse vesicle preparations in 10 ml editing buffer containing 0.6% (v/v) Nonidet-P40 for 30 min at 4°C and spin clear the lysate.

6. Equilibrate 1 ml of IgG sepharose beads (binding capacity: 1 mg/ml) in 10 ml IPP150.

7. Add 2 volumes of IPP150, the equilibrated IgG sepharose beads, and 5 U RNasin© to the cleared lysate and incubate for 1 h at 4°C. Transfer beads to a polypropylene column.

8. Wash IgG sepharose beads with 30 ml of IPP150 and equilibrate beads with 10 ml of TEV cleavage buffer.

9. Add 10 ml of TEV cleavage buffer, 5 μg/ml TEV protease, and 5 U/ml RNasin© to the IgG sepharose beads and incubate for 2 h at 16°C.

10. Equilibrate 0.5 ml of calmodulin affinity resin (binding capacity: 1 mg/ml) in 10 ml calmodulin binding buffer.

11. Collect the TEV eluate and add 2 volumes of calmodulin binding buffer, calmodulin affinity resin, and 5 U/ml RNasin©. Incubate for 1 h at 4°C.

12. Wash calmodulin affinity resin with 30 ml of calmodulin binding buffer.

13. Elute with calmodulin elution buffer in three 1 ml fractions.

14. Analyze an aliquot of each elution fraction in a 10% (w/v) SDS-containing polyacrylamide gel. Visualize by silver staining.

The TAP eluate was tested for its RNA editing activity using the precleaved RNA editing assays (17, 18). To analyze the TAP eluate on the protein level, protein bands were excised from SDS-containing polyacrylamide gels and "in gel"-digested with trypsin. Peptides were identified by mass spectrometry (MS). The MS spectra were analyzed using MS-Fit (19).

Apparent sedimentation coefficients (S-values) of the purified complexes were determined by isokinetic density centrifugation in linear glycerol gradients. To increase the detection limit, RNA editing complexes were radioactively labeled by protein iodination (20). In the presence of Chloramine-T, sodium iodide ($Na^{125}I$) is oxidized to molecular I_2 or ICl. Both molecules react specifically with tyrosines and to a lesser degree with histidines to form stable, covalent protein-^{125}I bonds.

1. Add 3.7 MBq $Na^{125}I$ (74 TBq/mmol) to 0.5 mg TAP eluate and transfer the sample to a fume hood (see Note 1). Start the reaction by adding 10 μl Chloramine-T (2.5 mg/ml) freshly dissolved in 50 mM sodium phosphate, pH 7.5. Incubate for 2 min at room temperature.

2. Quench the reaction by adding 20 μl of 3 mg/ml $Na_2S_2O_5$ freshly dissolved in 50 mM sodium phosphate, pH 7.5.

3. Separate radioactively labeled complexes in a 2.2 ml 0–40% (v/v) glycerol gradient for 2 h at $100,000 \times g$ at 4°C (Beckman TLS55 rotor).

4. Collect eleven 0.2 ml fractions from the top of the gradient.

5. Quantify gradient fractions by TCA precipitation followed by scintillation counting.

6. Determine the glycerol concentration of the gradient fractions by measuring the refractive index. To correlate a specific refractive index with an apparent S-value, marker molecules (5S rRNA, thyroglobin (19S), 23S rRNA, 30S and 50S ribosomal subunits) should be run on a separate gradient.

3.2. Surface Plasmon Resonance

Quantitative data for the interaction of mRNA, gRNA, and mRNA/gRNA hybrid molecules with ~20S editosomes can be derived from SPR experiments.

For a covalent attachment of the RNA to the SPR-microcuvette, 3′-oxidized RNA was bound to the surface of an amino silane microcuvette. In the presence of $NaIO_4$, RNA is oxidized at the 2′ and 3′ position of the 3′ terminal ribose ring by conversion of the hydroxyl groups to aldehyde groups. The aldehyde groups form Schiff's bases with the primary amines of the aminosilane surface, which can be reduced to secondary amines.

1. For the 3′ oxidation, incubate 10 μg of RNA with 20 mM $NaIO_4$ in oxidation buffer for 3 h at 4°C in the dark.

2. Purify oxidized RNA by gel filtration.

3. Insert an aminosilan-coated microcuvette into the IAsys instrument and add the oxidized RNA in 200 µl of coupling buffer. Incubate for 3 h at 27°C without stirring. Monitor the coupling reaction in real time.

4. Wash the cuvette 3× for 10 min with editing buffer.

The interaction of editing complexes with the RNA surface can be monitored in real time by SPR. All steps should be performed at 27°C with a stirrer frequency of 80 Hz. All buffers must be prewarmed to 27°C to avoid fluctuations in the resonant angle caused by a temperature shift.

1. Wash the cuvette with editing buffer until a stable baseline is reached.

2. Add a known editosome concentration (2–40 nM). An interaction of the RNA editing complexes with the RNA surface results in an increase of the resonant angle until saturation is reached. Typically, binding is completed within 1–2 min.

3. Add 3×200 µl editing buffer. Dissociation of editosomes from the RNA surface is achieved by dilution with excess of binding buffer. Dissociation results in a decrease of the resonant angle.

4. Regenerate the surface by adding 3×200 µl 6 M guanidinium-HCl in editing buffer. This step is required to remove remaining editosomes from the RNA surface.

5. Wash the cuvette 3× for 10 min with editing buffer and set the baseline. The cuvette is now ready for the next binding event.

6. Determine k_{diss} and k_{ass} values by plotting observed on rates $(k_{on(obs)})$ as a function of the complex concentration $(k_{on(obs)} = k_{ass} \times [complex] + k_{diss})$. Equilibrium dissociation constants (K_d) are calculated as $K_d = k_{diss}/k_{ass}$.

3.3. GraFix Fractionation

The RNA editing complexes isolated from the mitochondria of trypanosomes using the TAP-tag approach are a mixture of different assemblies (6). Thus, an additional purification step is favorable. To this end, we use gradient ultracentrifugation resulting in a separation of complexes dependent on their sedimentation characteristics. In addition, the complexes are further stabilized by a mild chemical fixation gradient during ultracentrifugation using the GraFix protocol (9). For visualization of the RNA editing complexes by EM, this approach was necessary (see Note 5).

1. Prepare glycerol or sucrose density gradients in a suitable buffer freshly (see Note 6). For RNA editing complexes, use low and high percentage buffers (see Note 2). Use a gradient forming device and keep the gradients at 4°C for about 1 h.

2. As an optional step for 13.2 ml gradient tubes, 0.5 ml of cushion buffer can be added prior to sample loading (see Note 7).

3. Carefully load the sample (up to 1.5 ml, depending on the sample concentration) on top of the gradient/cushion. Load at least 10–20 μg of protein on a 13.2 ml gradient. Overloading of the gradient should be avoided to prevent inter-particle crosslinking.

4. Spin, e.g., for 14 h at 247,000 × g and 4°C using a Sorvall TH-641 rotor.

5. Carefully fractionate the gradient from the bottom at 4°C using a needle introduced from the top of the gradient or alternatively use a piercing device (see Note 8). Fractions of 10 drops (corresponding to a volume of about 300 μl) are well suited for a 13.2 ml gradient.

3.4. Negative Staining and Cryo-negative Staining

Negative staining is a standard method to prepare macromolecular complexes for the visualization in an electron microscope; images are taken at room temperature. The approach uses various heavy metal stains such as uranyl salts. Due to the enhanced contrast also particles <100 kDa can be visualized depending on the shape. Cryo-negative staining combines negative staining with cryopreservation of the sample and a resolution in the sub-nanometer level can be obtained (21). Several different negative staining and cryo-negative staining protocols have been reported (22–24). We describe here the sandwich carbon procedure for imaging at room temperature and under cryogenic conditions. Both approaches were recently used to visualize endogenous RNA editing complexes isolated from trypanosomes (6).

1. Fill the sample (typically about 25 μl), the staining solution (about 100–200 μl), and a washing buffer (about 100–200 μl, optional step) into the wells of a black plastic or Teflon block. Keep the block throughout the sample preparation at 4°C (see Note 9). For uranyl salts, see also Note 3.

2. Carefully place a small piece (approximately 3 × 3 mm) of a continuous carbon film (that has previously been evaporated on mica in an indirect setup of the evaporating device) in the sample solution such that the carbon film detaches from the mica (Fig. 1a).

3. The incubation time depends on the specific sample and may vary between 15 s and 48 h. For editosomes, an adsorption time of 0.5–6 h was typically used.

4. For some samples, a washing step might be useful to increase the image contrast (Fig. 1b). This optional step can be done by transferring and incubating the carbon film for several seconds to few minutes in 100–200 μl of a washing buffer in

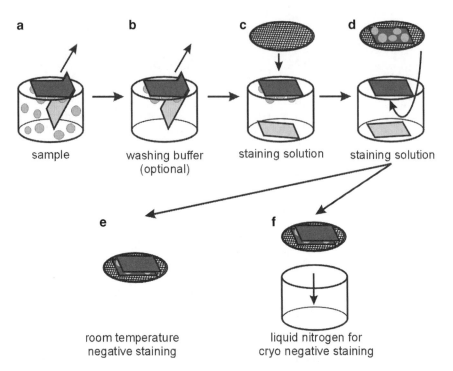

Fig. 1. Cryo-negative staining and negative staining. (**a**) Initially, the sample (particles are depicted as *gray spheres*) and the staining solution (two wells) are filled into wells of a black plastic or Teflon block. Optionally, wells can be filled with a washing buffer, if necessary. Most of the steps are identical for both room temperature negative staining and cryo-negative staining (**a–d**); the procedures differ only in the final step (**e** or **f**). A small piece of carbon film (*dark gray*) that was indirectly evaporated on a piece of mica (*light gray*) is placed into the sample solution. Thereby, the carbon film detaches from the mica except for a small area where the tweezers are touching it, and the macromolecules can adsorb for a defined period of time. (**b**) Optionally, the carbon film to which the particles have adsorbed can be transferred to a well filled with a washing buffer to improve the staining quality. This step can also be repeated, if necessary. (**c**) Subsequent to particle adsorption or the optional washing step, the carbon film is transferred to a well filled with a staining solution. Thereby, the mica completely detaches from the carbon film and falls down to the bottom of the well, while the carbon film with the particles facing towards the staining solution is floating on the surface of the staining solution. A copper EM grid (*top*) covered with a perforated carbon film is placed on top of the floating carbon film and the EM grid is carefully removed from the staining solution. Excess liquid is blotted from the side without destroying the carbon film. (**d**) Another piece of carbon film is placed into a second well filled with staining solution so that the carbon film is completely floating on the staining solution. The EM grid with the particles facing up is submerged underneath the second carbon film and lifted out of the staining solution to form the sandwich. Again, excess liquid is blotted. (**e**) For room temperature negative staining, the grid is dried and can then be stored at a dry place until imaging. (**f**) For cryo-negative staining, the EM grid is frozen in liquid nitrogen. Cryo grids must be stored and imaged under cryogenic conditions.

exactly the same way as carried out for the particle adsorption step (see Subheading 3.4, step 2).

5. Lift the carbon/mica out of the particle or optional washing solution, blot it carefully without touching the carbon film and transfer it to the staining solution (Fig. 1c). The carbon must completely detach from the mica. The particles face down towards the staining solution.

6. After about 2 min, the carbon film is lifted out of the staining solution by putting an EM grid covered with a perforated

carbon film on top of the floating carbon (Fig. 1c). Submerge the EM grid underneath the staining solution and carefully remove the EM grid out of the staining solution. Blot excess liquid from the side.

7. To form the carbon sandwich, float another piece of carbon film onto the surface of a second well filled with the staining solution so that the mica completely detaches from the carbon film (Fig. 1d). Submerge the EM grid with the particles facing up underneath the floating carbon film and lift the EM grid out of the solution to embed the particles in a layer of staining solution between the two carbon films. Again, blot the EM grid carefully from the side.

8. Let the EM grid air-dry for imaging at room temperature (Fig. 1e) or freeze the grid in liquid N_2 for cryo-negative staining (Fig. 1f). Once frozen, the grid must be kept under cryogenic conditions.

3.5. Unstained Cryopreparation

Unstained cryopreparations are obtained by freeze-plunging of the sample into liquid ethane (25). This results in the vitrification of the sample, i.e., the amorphous structure of the buffer is preserved and no crystalline ice is formed. Unstained cryoimaging can be performed for all samples of sufficient concentration and molecular mass (typically >250 kDa). In addition, the buffer must be compatible with vitrification. In particular, glycerol, sucrose, and high salt concentrations may interfere with the vitrification. In all such cases (e.g., GraFix fractions), the sample needs to be buffer exchanged to a suitable buffer.

1. Optional step in case substances with an adverse effect on the vitrification are present in the particle buffer: Buffer exchange the GraFix sample for a glycerol-free buffer using a Zeba spin column (see Note 10).

2. Fill 25–30 µl of sample into a well of the black plastic or Teflon block and proceed as described in Subheading 3.4, step 2. Keep the sample at 4°C.

3. The adsorption time needs to be adjusted according to the sample and may vary between 15 s and 48 h.

4. Place a copper EM grid covered with a perforated carbon film on top of the floating carbon film, submerge it underneath the surface, and carefully remove it out of the solution. Prior to usage, the EM grid can be glow discharged.

5. Mount the tweezers holding the EM grid in the freeze-plunger filled with liquid ethane (see also Note 4). Blot the grid carefully and plunge it into the liquid ethane (see Note 11). The EM grid is transferred and stored in liquid N_2 until imaging.

3.6. Electron Microscopy and Single-Particle Image Processing

The transmission electron microscope is used to image biological samples in the form of two-dimensional (2D) projection views. Due to the radiation sensitivity of biological material, imaging has to be conducted at low-dose conditions, and a further protection of the sample can be achieved by imaging at cryogenic temperatures (i.e., liquid N_2- or He-cooling) as compared to room temperature (26). To obtain the 3D information out of the data set of 2D projections, the particle views have to be subjected to single-particle image processing (7, 8). An overview of the steps is given in Fig. 2 and the single-particle image processing of the ~20S and ~35–40S RNA editing complexes isolated from *T. brucei* is summarized in Figs. 3 and 4, respectively.

1. Negatively stained grids can be transferred using standard room temperature holders into the microscope. For cryoimaging, the grid has to be mounted in a specialized side-entry cryoholder or a cryoloading system. During the mounting and transfer of the cryogrids, the sample has to be kept at cryogenic conditions (see Note 12).

2. For *de novo* structure determinations, a slow-scan CCD camera offers significant advantages compared to conventional photographic film; the latter usually is advantageous for high-resolution work (27). See Note 13. The magnification of the electron microscope has to be adjusted depending on, e.g., the size of the object, number of particles, pixel size of the detector, and desired resolution of the images. For CCD camera images, magnifications of about 50,000–275,000-fold and for photographic films, magnifications of 27,000–60,000-fold are typically used. Images can be taken untilted or at a tilt angle supported by the cryostage/holder combination. In particular, for the random conical tilt (RCT) technique (28), images are first taken at the selected tilt angle (e.g., 45°) and subsequently in an untilted mode at the same position to reduce the beam damage of the tilted images (as the tilted images are used for 3D calculation).

3. For photographic film only: Photographic film has to be digitized with a scanner of appropriate quality (e.g., Tango or Coolscan). Low quality images (e.g., poor contrast, drift, charging) can be discarded prior to this step.

4. Select individual particles from the raw EM images by manual or (semi-) automated procedures (see Note 14). Algorithms for manual and/or (semi-) automated particle selection are implemented in all major single-particle image processing software packages and individual programs are also available (e.g., (29–32)). Automated procedures can be combined with a postselection step in which all positions that were found by the software, but do not show a particle of desired quality are removed.

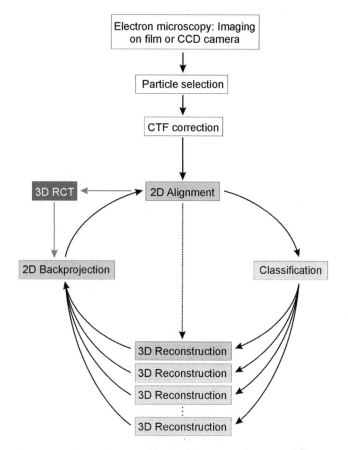

Fig. 2. Schematic representation of the individual steps performed during single-particle image processing. The initial steps (*white boxes*) comprise the imaging of the macromolecule, particle selection from the raw EM images, and correction for the CTF. The single-particle images are subsequently subjected to iterative rounds of single-particle image processing (labeled "2D Alignment," "Classification," "3D Reconstruction," and "2D Backprojection"). Initial 3D maps of a macromolecular assembly can be calculated using the RCT (labeled "3D RCT") or angular reconstitution approach. During the refinement phase, the alignment process itself can be used for grouping of similar views in alignment averages and these can be used for 3D calculation (*dashed arrow*). Competitive multireference alignment can be performed for samples of dynamic macromolecular assemblies characterized by the presence of multiple 3D structural states (labeled "3D Reconstruction" in the bottom part of the circle). Thereby, the individual 3D maps are backprojected into the 2D space (labeled "2D Backprojection"; in case of competitive multireference alignment, these 2D backprojections derive from different 3D maps) and each single-particle image is competitively aligned to these reference projections.

5. Correct for the contrast-transfer-function (CTF) imposed on the raw EM images (see Note 15). Programs for this are included in all major software packages as well as in individual programs (e.g., (29–31, 33)). This step also offers the possibility to identify low quality images.

6. The particle images can now be used for iterative rounds of 2D and 3D image processing (Fig. 2). This is typically done

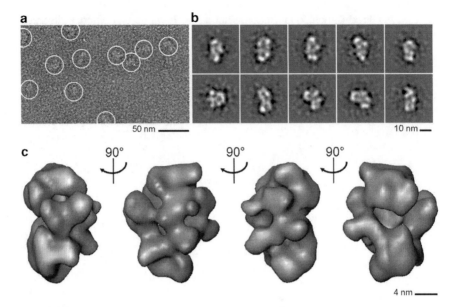

Fig. 3. Single-particle electron microscopy of the ~20S editosome of *Trypanosoma brucei* (6). (**a**) Raw electron micro-scopical image of negatively stained ~20S RNA editing complexes. Individual particles are encircled. (**b**) Class averages of the ~20S editosome as obtained upon 2D single-particle image processing. Some of the class averages represent different structural subtypes of the ~20S complexes and thus belong to different 3D maps. (**c**) 3D map of the ~20S edito-some using cryo-negatively stained images. Subsequent to separation of the data set into structural subtypes by com-petitive multireference alignment, the 3D structure was determined from a subset of images belonging to one structural subtype of the ~20S editosome. 3D views were slightly rotated with respect to the published orientations (6).

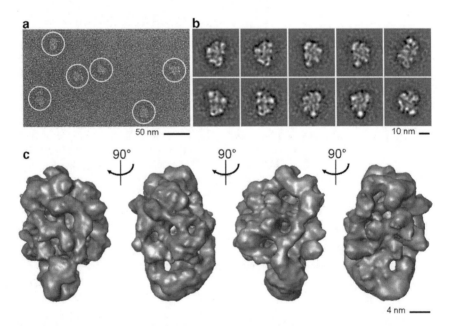

Fig. 4. Single-particle electron microscopy of the ~35–40S editosome of *Trypanosomal brucei* (6). (**a**) Raw electron micro-scopical image of negatively stained ~35–40S RNA editing complexes. Individual particles are encircled. (**b**) Class averages of the ~35–40S editosome as obtained upon 2D single-particle image processing. Some of the class averages represent different structural subtypes of the ~35–40S complexes and thus belong to different 3D maps. (**c**) 3D map of the ~35–40S editosome using cryo-negatively stained images. Subsequent to separation of the data set into structural subtypes by competitive multireference alignment, the 3D structure was determined from a subset of images belonging to one struc-tural subtype of the ~35–40S editosome. 3D views were slightly rotated with respect to the published orientations (6).

in one of the major software packages such as, e.g., IMAGIC (29), SPIDER (30), EMAN (31), XMIPP (34), or FREALIGN (35) (see Note 16). One round of 2D image processing typically comprises of an alignment step and a classification step. During the alignment, the particle images are aligned towards references to bring similar views into register (see Note 17). These aligned images can subsequently be used to group similar views into classes in order to increase the signal-to-noise-ratio. Alignment and classification are iteratively repeated until the result is stable.

7. The class or alignment averages are then used to calculate the 3D map of the macromolecular assembly. For this, all programs listed above can be used. Several different techniques exist including, e.g., RCT (28), angular reconstitution (36), and projection matching (37) (see Note 18).

8. For a refinement of the initial model (e.g., a low-resolution 3D map determined using the RCT or angular reconstitution approach), the Subheading 3.6, steps 6 and 7 are repeated until the result is stable and no improvement of the structure is observed. 2D backprojections of the 3D map are used as an alignment reference during the refinement of the 3D map. For dynamic macromolecular assemblies, several 3D structural maps representing different compositional and/or conformational states can be determined using the RCT technique. These different 3D maps can be used to generate sets of 2D backprojections for a competitive multireference alignment. Each single-particle image is thereby competitively aligned to these reference projections, and the data set is separated into subsets representing individual structural subtypes. These individual structural subtypes are refined separately. Typically, during the refinement steps, a large number of references is used to cover all possible angular directions. This makes more advanced schemes such as the corrims (38) necessary to speedup the calculation process.

9. Finally, the 3D map is visualized as sections and/or surface views in a 3D viewer and can be annotated by, e.g., fitting of substructures, labeling, recognition of structural elements, and other techniques.

4. Notes

1. ^{125}I emits gamma rays with a maximal energy of 0.035 MeV and requires lead shielding. Unbound iodine is volatile and must be handled under a fume hood.

2. We recommend using fresh EM-grade solutions of 25% (v/v) glutaraldehyde since the activity of the crosslinker may decrease over time. Glutaraldehyde and other crosslinkers are highly toxic and appropriate safety precautions according to the manufacturer should be followed.

3. Prepare the uranyl formate solution always freshly and keep the solution cool and in the dark. Uranyl salts are radioactive and toxic and thus appropriate safety precautions should be followed.

4. Liquid N_2 and liquid ethane can cause severe burns. Note the asphyxiation hazard. Ethane is extremely flammable and forms explosive mixtures with air. Appropriate protection is required.

5. Depending on the particle, GraFix can have a number of advantages: (1) significant reduction of particle disintegration and aggregation, (2) improved adsorption towards the carbon support film, (3) increased image contrast, and (4) increase in angular diversity.

6. As the gradient contains a low amount of glutaraldehyde, only buffer reagents compatible with the fixation reagent can be used. Glutaraldehyde, for example, reacts with primary amines (39, 40), and thus all gradient buffers must be free of primary amines.

7. Some purification protocols result in the presence of primary amine compounds in the sample in addition to the protein or RNA/protein complex of interest. These include, e.g., elution peptides contaminating polypeptides at high concentration and Tris-OH. As a direct contact of such a sample with the crosslinker has to be avoided, a cushion free of primary amines and free of glutaraldehyde is used before loading the sample.

8. We highly recommend fractionating the gradient from the bottom since low molecular mass substances such as peptides and detergents are enriched in the top fractions of the gradient. These low molecular mass substances may interfere with the image contrast of the EM specimen and may cause staining artifacts. Using a needle/tube system connected to a peristaltic pump, mixing of the gradient, e.g., during the insertion of the needle has to be avoided.

9. During long-term incubation, the block has to be kept at 4°C and condensation of water vapor on the sample must be avoided. This can be accomplished by placing the block into a dry, precooled Petri dish, mounting the lid and storing the block/Petri dish in a cold room. Even during short-term incubation, the block must be kept at 4°C and condensation of water vapor must be avoided. However, it is usually sufficient to place the block in an ice-cooled Petri dish.

10. As the recovery of the sample upon buffer exchange may vary with the properties of the selected particle (for large multi-megadalton assemblies in the range of about 20–95%), we recommend to prepare a negative staining room temperature EM grid to adapt the incubation time for the unstained cryo-preparation accordingly.

11. The blotting properties (i.e., time, pressure, one-sided vs. two-sided, frequency, etc.) as well as the environmental properties (temperature and humidity) have to be adapted to the specific sample under investigation. Assessment of the vitrification quality must be done by using the electron cryomicroscope under cryoconditions.

12. Mounting and transfer of room temperature EM grids is straightforward. Significant warming and ice contamination of cryo-EM grids during the mounting and transfer under cryogenic conditions must be avoided.

13. Slow-scan CCD cameras offer an increased phase transmission and spectral signal-to-noise ratio (27), which are important factors in the setup of a novel structure. Images can be recorded in tile mode – i.e., with slight overlap of the images – and stitched to larger images to compensate for the smaller imaging area of the CCD camera. In contrast, photographic film shows superior quality in the very high resolution range.

14. Manual selection of particles is more time-intensive compared to (semi-) automated selection procedures, but offers the opportunity to get an overview about the typical views and the quality of the imaged particles (e.g., homogeneity, aggregation, disintegration). In particular for particles with so-far unknown structure, manual selection is advantageous. For the selection of larger data sets of particles with known structure, automated particle selection can be performed using "Boxer" (EMAN software package (31)) or "Signature" (32). Manual postselection is often recommended to remove low quality images (e.g., aggregates, broken particles, ice contamination). In any of the methods, the particles should be picked centrally, i.e., a displacement of the particle in direction of the edges should be avoided. Particles are extracted from the raw images in a pixel frame larger than the actual maximum dimensions of the visualized particles. In general, a pixel frame in which the particles amount to about two thirds of the frame is recommended. Depending on the particle dimensions and the pixel size, typical values of the pixel frame are 64×64 to 512×512 pixels.

15. Computationally, raw EM images should be corrected for the defocus and twofold astigmatism, and can also be corrected for the experimental B factor, if desired. Low quality images due to drift or charging can be easily identified in the 2D power spectra as truncations of the Thon rings.

16. File format type and definition of the coordinate system including 3D angles may vary between the individual programs and thus need to be checked carefully when switching from one software package to the other.

17. There are several ways to generate reference images for the alignment. In case of a particle with an unknown structure, selected single-particle views or the sum of all images (i.e., a featureless "blob") can be used as a first reference. Also, views from a related particle (e.g., with minor differences in composition) can be used in the first alignment. For subsequent rounds, class or alignment averages can be used (see below). Once a 3D model of the analyzed or a related (e.g., with minor differences in composition) particle is available, this can be used to generate 2D projections by backprojecting the 3D map into 2D projections. For averaging, either a statistical classification approach such as multivariate statistical classification or – in particular in the advanced steps of the refinement – the alignment itself is used. The former technique generates so-called class averages, the latter alignment averages.

18. All of the different 3D reconstruction techniques are implemented in one or more of the software packages listed in Subheading 2.7, item 8. The selection of the approach primarily depends on the stage of the project. RCT (28) and angular reconstitution (36) can be used to determine the structure of a particle whose structure was so-far unknown, i.e., in a *de novo* approach. For refinement of an available model, angular reconstitution and projection matching (37) can be used. RCT structures are typically limited to the low resolution range and suffer from a missing cone, but in contrast to the other techniques can determine the handedness of the complex. Refinement of an RCT structure is possible by both angular reconstitution and projection matching. Angular reconstitution and projection matching can reconstruct 3D maps to the subnanometer level.

Acknowledgments

MMG and BS are supported by a grant from the Danish Center for Scientific Computing (DCSC). HS is supported by a grant of the Bundesministerium für Bildung und Forschung (BMBF) and a European "3D Repertoire" grant. HUG is supported as an International Scholar of the Howard Hughes Medical Institute (HHMI) and by the German Research Foundation (DFG).

References

1. Alberts, B. (1998) The cell as a collection of protein machines: preparing the next generation of molecular biologists *Cell* **92**, 291–4.

2. Madison-Antenucci, S., Grams, J., and Hajduk, S. L. (2002) Editing machines: the complexities of trypanosome RNA editing *Cell* **108**, 435–8.

3. Stuart, K. D., Schnaufer, A., Ernst, N. L., and Panigrahi, A. K. (2005) Complex management: RNA editing in trypanosomes *Trends Biochem Sci* **30**, 97–105.

4. Stark, H., and Lührmann, R. (2006) Cryo-electron microscopy of spliceosomal components *Annu Rev Biophys Biomol Struct* **35**, 435–57.

5. Leschziner, A. E., and Nogales, E. (2007) Visualizing flexibility at molecular resolution: analysis of heterogeneity in single-particle electron microscopy reconstructions *Annu Rev Biophys Biomol Struct* **36**, 43–62.

6. Golas, M. M., Böhm, C., Sander, B., Effenberger, K., Brecht, M., Stark, H., and Göringer, H. U. (2009) Snapshots of the RNA editing machine in trypanosomes captured at different assembly stages *in vivo* *EMBO J* **28**, 766–78.

7. van Heel, M., Gowen, B., Matadeen, R., Orlova, E. V., Finn, R., Pape, T., Cohen, D., Stark, H., Schmidt, R., Schatz, M., and Patwardhan, A. (2000) Single-particle electron cryo-microscopy: towards atomic resolution *Q Rev Biophys* **33**, 307–69.

8. Frank, J. (2002) Single-particle imaging of macromolecules by cryo-electron microscopy *Annu Rev Biophys Biomol Struct* **31**, 303–19.

9. Kastner, B., Fischer, N., Golas, M. M., Sander, B., Dube, P., Boehringer, D., Hartmuth, K., Deckert, J., Hauer, F., Wolf, E., Uchtenhagen, H., Urlaub, H., Herzog, F., Peters, J. M., Poerschke, D., Lührmann, R., and Stark, H. (2008) GraFix: sample preparation for single-particle electron cryomicroscopy *Nat Methods* **5**, 53–5.

10. Brun, R., and Schönenberger, M. (1979) Cultivation and *in vitro* cloning of procyclic culture forms of *Trypanosoma brucei* in a semi-defined medium *Acta Trop* **36**, 289–92.

11. Wirtz, E., Leal, S., Ochatt, C., and Cross, G. A. (1999) A tightly regulated inducible expression system for conditional gene knock-outs and dominant-negative genetics in Trypanosoma brucei *Mol Biochem Parasitol* **99**, 89–101.

12. Rigaut, G., Shevchenko, A., Rutz, B., Wilm, M., Mann, M., and Seraphin, B. (1999) A generic protein purification method for protein complex characterization and proteome exploration *Nat Biotechnol* **17**, 1030–2.

13. Brecht, M., Niemann, M., Schlüter, E., Müller, U. F., Stuart, K., and Göringer, H. U. (2005) TbMP42, a protein component of the RNA editing complex in African trypanosomes, has endo-exoribonuclease activity *Mol Cell* **17**, 621–30.

14. Panigrahi, A. K., Schnaufer, A., Carmean, N., Igo, R. P., Jr., Gygi, S. P., Ernst, N. L., Palazzo, S. S., Weston, D. S., Aebersold, R., Salavati, R., and Stuart, K. D. (2001) Four related proteins of the Trypanosoma brucei RNA editing complex *Mol Cell Biol* **21**, 6833–40.

15. Cross, G. A. (1975) Identification, purification and properties of clone-specific glycoprotein antigens constituting the surface coat of Trypanosoma brucei *Parasitology* **71**, 393–417.

16. Hauser, R., Pypaert, M., Hausler, T., Horn, E. K., and Schneider, A. (1996) *In vitro* import of proteins into mitochondria of Trypanosoma brucei and Leishmania tarentolae *J Cell Sci* **109 (Pt 2)**, 517–23.

17. Igo, R. P., Jr., Palazzo, S. S., Burgess, M. L., Panigrahi, A. K., and Stuart, K. (2000) Uridylate addition and RNA ligation contribute to the specificity of kinetoplastid insertion RNA editing *Mol Cell Biol* **20**, 8447–57.

18. Igo, R. P., Jr., Weston, D. S., Ernst, N. L., Panigrahi, A. K., Salavati, R., and Stuart, K. (2002) Role of uridylate-specific exoribonuclease activity in Trypanosoma brucei RNA editing *Eukaryot Cell* **1**, 112–8.

19. Clauser, K. R., Baker, P., and Burlingame, A. L. (1999) Role of accurate mass measurement (+/- 10 ppm) in protein identification strategies employing MS or MS/MS and database searching *Anal Chem* **71**, 2871–82.

20. Hunter, W. M., and Greenwood, F. C. (1962) Preparation of iodine-131 labelled human growth hormone of high specific activity *Nature* **194**, 495–6.

21. Golas, M. M., Sander, B., Will, C. L., Lührmann, R., and Stark, H. (2003) Molecular architecture of the multiprotein splicing factor SF3b *Science* **300**, 980–4.

22. Harris, J. R. (2007) Negative staining of thinly spread biological samples *Methods Mol Biol* **369**, 107–42.

23. Adrian, M., Dubochet, J., Fuller, S. D., and Harris, J. R. (1998) Cryo-negative staining *Micron* **29**, 145–60.

24. Golas, M. M., Sander, B., Will, C. L., Lührmann, R., and Stark, H. (2005) Major conformational change in the complex SF3b upon integration into the spliceosomal U11/U12 di-snRNP as revealed by electron cryomicroscopy *Mol Cell* **17**, 869–83.

25. Adrian, M., Dubochet, J., Lepault, J., and McDowall, A. W. (1984) Cryo-electron microscopy of viruses *Nature* **308**, 32–6.

26. Chiu, W., Downing, K. H., Dubochet, J., Glaeser, R. M., Heide, H. G., Knapek, E., Kopf, D. A., Lamvik, M. K., Lepault, J., Robertson, J. D., Zeitler, E., and Zemlin, F. (1986) Cryoprotection in electron microscopy *J Microsc* **141**, 385–91.

27. Sander, B., Golas, M. M., and Stark, H. (2005) Advantages of CCD detectors for *de novo* three-dimensional structure determination in single-particle electron microscopy *J Struct Biol* **151**, 92–105.

28. Radermacher, M. (1988) Three-dimensional reconstruction of single particles from random and nonrandom tilt series *J Electron Microsc Tech* **9**, 359–94.

29. van Heel, M., Harauz, G., Orlova, E. V., Schmidt, R., and Schatz, M. (1996) A new generation of the IMAGIC image processing system *J Struct Biol* **116**, 17–24.

30. Shaikh, T. R., Gao, H., Baxter, W. T., Asturias, F. J., Boisset, N., Leith, A., and Frank, J. (2008) SPIDER image processing for single-particle reconstruction of biological macromolecules from electron micrographs *Nat Protoc* **3**, 1941–74.

31. Tang, G., Peng, L., Baldwin, P. R., Mann, D. S., Jiang, W., Rees, I., and Ludtke, S. J. (2007) EMAN2: an extensible image processing suite for electron microscopy *J Struct Biol* **157**, 38–46.

32. Chen, J. Z., and Grigorieff, N. (2007) SIGNATURE: a single-particle selection system for molecular electron microscopy *J Struct Biol* **157**, 168–73.

33. Sander, B., Golas, M. M., and Stark, H. (2003) Automatic CTF correction for single particles based upon multivariate statistical analysis of individual power spectra *J Struct Biol* **142**, 392–401.

34. Sorzano, C. O., Marabini, R., Velazquez-Muriel, J., Bilbao-Castro, J. R., Scheres, S. H., Carazo, J. M., and Pascual-Montano, A. (2004) XMIPP: a new generation of an open-source image processing package for electron microscopy *J Struct Biol* **148**, 194–204.

35. Grigorieff, N. (2007) FREALIGN: high-resolution refinement of single particle structures *J Struct Biol* **157**, 117–25.

36. van Heel, M. (1987) Angular reconstitution: a posteriori assignment of projection directions for 3D reconstruction *Ultramicroscopy* **21**, 111–23.

37. Penczek, P. A., Grassucci, R. A., and Frank, J. (1994) The ribosome at improved resolution: new techniques for merging and orientation refinement in 3D cryo-electron microscopy of biological particles *Ultramicroscopy* **53**, 251–70.

38. Sander, B., Golas, M. M., and Stark, H. (2003) Corrim-based alignment for improved speed in single-particle image processing *J Struct Biol* **143**, 219–28.

39. Hopwood, D. (1972) Theoretical and practical aspects of glutaraldehyde fixation *Histochem J* **4**, 267–303.

40. Migneault, I., Dartiguenave, C., Bertrand, M. J., and Waldron, K. C. (2004) Glutaraldehyde: behavior in aqueous solution, reaction with proteins, and application to enzyme crosslinking *Biotechniques* **37**, 790–6, 8–802.

Chapter 2

iCODA: RNAi-Based Inducible Knock-In System in *Trypanosoma brucei*

Gene-Errol Ringpis, Richard H. Lathrop, and Ruslan Aphasizhev

Abstract

In vivo mutational analysis is often required to characterize enzymes that function as subunits of the U-insertion/deletion RNA editing core complex (RECC) in mitochondria of *Trypanosoma brucei*. The mutations may skew phenotypic manifestation of a dominant negative overexpression if complex association is disrupted. Conditional knockouts and knock-ins of essential mitochondrial genes are time consuming and restricted to the bloodstream form parasites, thus limiting biochemical analysis. We have combined CODA (computationally optimized DNA assembly) technology with RNA interference to develop an iCODA inducible knock-in system for expeditious phenotype assessment and affinity purification of the RECC bearing a mutant subunit. For functional knock-in, the gene region targeted by RNAi is replaced with a synthetic sequence bearing at least one silent mutation per 12 contiguous base pairs. Upon co-expression of the double-stranded RNA targeting the endogenous transcript and modified mRNA in a stable cell line, the endogenous mRNA is destroyed and the cell survives on the RNAi-resistant transcript encoding the same polypeptide. In this chapter, we describe the generation of procyclic (insect) transgenic cell lines, RNAi rescue, complex purification, and validation methods for RNA editing TUTase 2 (RET2). These methods should be readily applicable for any gene in *T. brucei*.

Key words: *Trypanosoma*, Mitochondria, RNA editing, RNAi, TUTase

1. Introduction

Proteomic studies of trypanosomal mitochondrial RNA editing complexes, including the RECC (1–3) and the guide RNA binding complex (4), have been expedited by efficient constitutive and inducible overexpression systems. Introduction of a C-terminal affinity TAP tag (5), composed of protein A, calmodulin binding peptide, and a tobacco etch virus (TEV) protease cleavage site, enabled isolation and mass spectrometry analysis of these and other complexes involved in mitochondrial RNA

Ruslan Aphasizhev (ed.), *RNA and DNA Editing: Methods and Protocols*, Methods in Molecular Biology, vol. 718,
DOI 10.1007/978-1-61779-018-8_2, © Springer Science+Business Media, LLC 2011

processing (2, 6). Because *Trypanosoma brucei* is a diploid asexually reproducing organism, inducible knockouts of essential genes require sequential generation of three clonal cell lines and as many selective markers (7); a fourth marker is required for a knock-in construction. For reasons that are not clear, successful knockouts and knock-ins of essential mitochondrial genes (8, 9) have been achieved only in blood stream (BF) parasites which are deficient in oxidative phosphorylation and grow to a low cell density (~10^6 cells/mL). These technical restrictions made RNA interference (10) a method of choice for gene silencing in BF and procyclic (PF) parasites, which have an actively respiring mitochondrion and can be cultured at higher cell density (~10^7 cells/mL).

Transgenic cell lines have been generated for both BF and PF *T. brucei* to allow tetracycline-regulated (11) ectopic expression of double-stranded (ds) RNA driven by either T7 RNA polymerase or RNA polymerase I promoters (12). Typically, the RNAi cassette is designed toward protein-coding regions, although achieving sufficient knockdown may require empirical selection among several gene fragments (13) or inclusion of untranslated regions (UTRs) (14). Targeting unique UTRs for RNAi also has been used as a knock-in method. In this strategy, RNAi targets exclusively 5'- or 3'-UTR triggering degradation of the endogenous mRNA while the gene of interest flanked by heterologous UTRs is expressed (15). Limitations of this approach include a requirement for precise UTR mapping to avoid knockdown of closely spaced adjacent mRNAs and UTRs that are too short for effective RNAi knockdown.

Although biochemical and structural analyses of recombinant editing enzymes were instrumental in defining catalytic residues and domain organization (16, 17), functional studies would greatly benefit from an *in vivo* mutagenesis system which also allows *in vitro* analysis of purified editing complexes. We have developed an iCODA *in vivo* complementation approach for PF parasites based on tetracycline-inducible co-expression of a dsRNA and a synthetic gene which encodes the same polypeptide as the one targeted by RNAi. A fragment corresponding to the RNAi-targeted region is assembled from overlapping DNA oligonucleotides using CODA technology (18) with at least one silent mutation per 12 bp. These mutations are designed based on a genome-wide analysis of codon bias and codon context to minimize effects on translation. Transcription of the synthetic gene produces an RNAi-resistant mRNA as the only source for the protein of interest. Consequentially, the cell survives unless a mutation disrupting catalysis or complex association is introduced into the synthetic gene. Introduction of the TAP tag allows for protein complex purification and downstream biochemical analysis (Fig. 1).

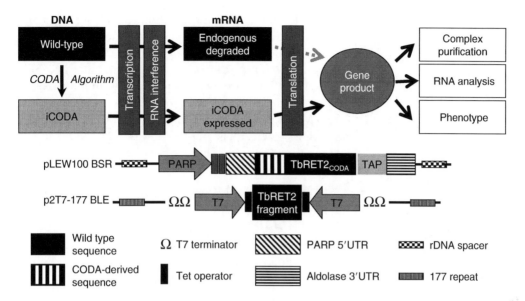

Fig. 1. iCODA knockdown/knock-in strategy. Silent mutations (at least one per 12 bp, and often more) were introduced into the RET2 gene to minimize potential effects on translation (18) and to prevent transcript targeting by the RNAi machinery. The expression of both RET2-iCODA protein and RET2 RNAi cassette is controlled by tet-operators positioned downstream of a procyclic acidic repetitive protein promoter (PARP), which is recognized by RNA polymerase I and T7 RNA polymerase promoter, respectively.

The iCODA technology has been applied to studies of RNA editing terminal uridylyl transferase 2 (RET2), an integral RECC component responsible for the U-insertion mRNA editing activity (19, 20). RET2 is the only nonredundant enzyme within the RECC and is essential for parasite viability in both PF (19) and BF (17). In addition, the iCODA technology has been useful in validating RNAi knockdown specificity for partial 3′-UTR targeting in the case of MEAT1 TUTase (14). Here, we present methods for design of an RNAi resistant gene, generation and validation of iCODA/RNAi cell lines, and complex purification.

2. Materials

2.1. Genetic Constructs

1. p2T7-177-BLE – RNAi construct with opposing T7 RNA polymerase promoters and phleomycin resistance (21).

2. pLEW100-TAP-BSR – Protein expression construct with blasticidin resistance (available from author's laboratory).

2.2. Cell Culture and Selection of Clonal Cell Lines

1. 29-13 strain of procyclic *T. brucei* (12).

2. SDM-79 (semi-defined) medium: dissolve 7 g S-MEM (Invitrogen), 2 g Medium 199 (Invitrogen), 1 g d-glucose, 8 g HEPES, 5 g MOPS, 2 g $NaHCO_3$, 100 mg sodium pyruvate,

200 mg L-alanine, 100 mg L-arginine, 300 mg L-glutamine, 70 mg L-methionine, 80 mg L-phenylalanine, 600 mg L-proline, 60 mg L-serine, 160 mg L-taurine, 350 mg L-threonine, 200 mg L-tyrosine, 10 mg adenosine, 10 mg guanosine, 50 mg glucosamine-HCl, 4 mg folic acid, 2 mg p-aminobenzoic acid, 200 µg biotin, 8 mL MEM amino acids (50×) (Invitrogen), 6 mL MEM nonessential amino acids (100×) (Invitrogen) in 850 mL of water. Adjust pH to 7.3 with 5 M NaOH and bring volume to 900 mL. Sterilize by filtration and store at +4°C up to 2 weeks.

3. 29-13 Medium: SDM-79 medium supplemented with 50 µg/mL of Geneticin (G418), 50 µg/mL of Hygromycin, 10 µg/mL of hemin (EMD Chemicals), and 10% heat inactivated fetal bovine serum (FBS, see Note 1).

4. Limiting dilution (LD) medium: same as 29-13 medium with appropriate additional drug(s) and 20% FBS.

5. Humidified incubator set at 27°C with 5% CO_2.

2.3. Electroporation

1. Not I restriction endonuclease.

2. Cytomix: 120 mM KCl, 0.15 mM $CaCl_2$, 10 mM potassium phosphate, 25 mM HEPES, 2 mM EDTA, 5 mM $MgCl_2$, pH 7.6.

3. Phosphate sucrose buffer: 277 mM sucrose, 1 mM $MgCl_2$, 7 mM potassium phosphate, pH 7.4.

4. EM buffer: 3:1 mixture of Cytomix and phosphate sucrose buffer.

5. Bio-Rad Gene Pulser.

6. 0.4-cm electroporation cuvettes.

2.4. Tandem Affinity Purification

1. Extraction buffer (EB): 50 mM Tris–HCl pH 7.6, 150 mM KCl, 2 mM EDTA. Stock of Tris–HCl is prepared as 1 M solution with pH 7.6 at 20°C.

2. IgG binding buffer (IBB): 25 mM Tris–HCl pH 7.6, 150 mM KCl, 1 mM EDTA, 0.1% NP-40.

3. TEV cleavage buffer (TCB): IBB supplemented with 1 mM DTT (see Note 2).

4. Calmodulin binding buffer (CBB): 25 mM Tris–HCl pH 7.6, 150 mM KCl, 0.1% NP-40, 10 mM β-mercaptoethanol, 1 mM magnesium acetate, 1 mM imidazole, 2 mM $CaCl_2$.

5. Calmodulin elution buffer 1 (CEB1) for activity assays: 25 mM Tris–HCl pH 7.6, 150 mM KCl, 3 mM EGTA, 1 mM EDTA, 0.1% NP40, 10% glycerol.

6. Calmodulin elution buffer 2 (CEB2) for mass spectrometry analysis of complexes: 25 mM Tris–HCl pH 7.6, 50 mM KCl, 3 mM EGTA, 1 mM EDTA, 0.1% NP40.

7. Disposable polystyrene columns (Pierce, product # 29920).

8. TEV protease (Invitrogen).

9. Sypro Ruby gel stain (Invitrogen).

10. Complete protease inhibitor (Roche). Dissolve one tablet in 1 mL of water.

3. Methods

3.1. Genetic Constructs Design

The choice of ~400 bp-long DNA fragment for the RNAi cassette is based on the RNAit algorithm (http://trypanofan.path.cam. ac.uk/software/RNAit.html) that uses BLAST searches to minimize off-targeting (22). The p2T7-177 genetic construct (partially diagrammed in Fig. 1) is a standard vehicle for inducible expression of dsRNA fragments in trypanosomes (21). The dual opposing T7 promoter architecture of this vector allows a one-step cloning of the target sequence or multiple gene knockdowns by cloning several target sequences. Integration into the mini-chromosome 177-bp repeat region provides for low baseline transcription.

The pLEW100 vector (23), partially diagrammed in Fig. 1, is widely used for stable expression in *T. brucei* via integration into the rRNA spacer. Procyclic acidic repetitive protein (PARP) promoter-driven expression regulated by *tet*-operators allows for inducible expression in PF parasites. Alternatively, if work in BF parasites is planned, the pLEW100v5-BSD vector (http://tryps. rockefeller.edu) may be a better alternative. In this construct, expression is driven by the rRNA promoter that is active in both PF and BF *T. brucei*. While overexpression is often performed in the pLEW79-based mhTAP vector (24), pLEW100 variants provide a balance of efficient integration and tightly regulated expression which is better suited for iCODA.

The synthetic gene replacement construct is designed to (a) be identical in amino acid sequence to the wild-type gene except for deliberately introduced mutations; (b) differ from the wild-type gene DNA sequence by at least 1 bp in any contiguous 12 bp; (c) avoid off-target cross-hybridization, which is known to be a source of mutagenesis failure (25); (d) respect known or hypothesized codon usage (26) and codon pair (27–29) preferences; (e) avoid any undesired DNA sequences, e.g., restriction sites used in cloning; and (f) be easily assembled from purchased synthetic oligonucleotides by polymerase extension.

These goals are all accomplished by making silent (synonymous) codon substitutions in the designed synthetic gene replacement construct. For this purpose, we used the computationally optimized DNA assembly technology (CODA, (18)) to design the gene assembly and mfold (30) to identify undesired RNA secondary structures. However, any modern gene design software that accepts sequence constraints and removes undesired RNA secondary structures may be used instead of CODA. Conceptually, the CODA system works by beginning with (a) a desired amino acid sequence, (b) a candidate DNA sequence, and (c) a set of constraints to satisfy. It then iteratively introduces silent codon substitutions into the candidate DNA sequence until all constraints are satisfied. Finally, it outputs (a) a list of DNA oligonucleotide sequences to purchase and (b) a set of instructions for combining the purchased oligonucleotides by polymerase extension. The actual CODA implementation involves more technical details; see Larsen et al. for further explanation (18).

1. Define the desired synthetic replacement sequence. To minimize the DNA synthesis and assembly burden, an approximately 400 bp gene fragment is recommended for iCODA.

2. Prepare an initial candidate DNA sequence by replacing every codon in the sequence from step 1 by the most prevalent (most highly used) codon for that amino acid in the host organism (26), unless the wild-type DNA sequence also uses the most prevalent codon there, in which case choose the second most prevalent codon. The result of this step is a DNA sequence that is identical in encoded amino acid sequence to step 1 and differs from the wild-type DNA sequence at every codon (except for methionine and tryptophan, which each have only one codon).

3. Prepare a list of desired DNA sequence constraints, which includes any required DNA subsequences while prohibiting (a) any four consecutive codons from the wild-type gene; (b) restriction sites used in cloning from occurring anywhere in the coding region of the gene; (c) low-usage codons or undesired codon pairs; and (d) any other DNA sequences to exclude. Prohibiting any four consecutive codons from the wild-type gene guarantees that the desired synthetic gene replacement construct has no more, and often fewer, than 11 contiguous base pairs that are identical to the wild-type gene DNA sequence.

4. Compare the current candidate DNA sequence to the list of sequence constraints from step 3. If any constraint violations are found, introduce silent (synonymous) codon substitutions into the candidate DNA sequence to eliminate them while avoiding low-usage codons (26) and go to step 4.

5. Analyze the current candidate DNA sequence for undesired RNA secondary structures, such as hairpins, cross-hybridizations, etc. (18, 25). If any undesired RNA secondary structures are found, introduce silent (synonymous) codon substitutions

into the candidate DNA sequence to eliminate them while avoiding low-usage codons (26), and go to step 4.

6. Divide the DNA sequence of step 5 into DNA oligonucleotide sequences that can be purchased commercially and assemble them by polymerase extension as described in Larsen et al. (18).

3.2. Transfection of Genetic Constructs into 29-13 T. brucei by Electroporation

The following protocol may be used for transfection with a single plasmid or co-transfection with two plasmids (see Note 3). All materials and solutions that will come in contact with cells should be kept on ice. Sterile technique must be employed when handling cells.

3.2.1. Preparation of Plasmids and Cell Lines

1. From a mid-logarithmic culture maintained at ~5×10^6 cells/mL in 29-13 medium, seed $1–2 \times 10^6$ cells/mL to achieve a cell density of $4–5 \times 10^6$ cells/mL on the following day. Assume that the division time is ~12 h and 1.4×10^7 cells per transfection are required. Incubate 25–30 mL of suspension culture at 27°C with mild agitation (~100 rpm) in a 75 cm^2 cell culture flask in vertical position.

2. Digest 10–12 µg of plasmid DNA purified with QIAprep kit (Qiagen) with 5 units of Not I restriction endonuclease in a 200-µL reaction volume at 37°C overnight.

3. Precipitate the DNA with 500 µL of ethanol for 20 min in dry ice. Do not add extra salt or extract with organic solvents. Wash the pellet with 70% ethanol, air-dry in the laminar hood, and resuspend in 100 µL of sterile water (50 µL for co-transfection).

3.2.2. Electroporation

1. When the cell count reaches ~5×10^6/mL, transfer cells into a conical tube.

2. Using a fixed-angled rotor, centrifuge the cells $2,000 \times g$ at 4°C for 5 min. Aspirate the supernatant completely.

3. Fill the tube to 50% volume with EM buffer and gently resuspend cells with pipetting. Repeat centrifugation and aspiration.

4. Resuspend cells in EM buffer at 3×10^7/mL.

5. Transfer the DNA solution into a chilled 4-mm electroporation cuvette. Use sterile water for the control transfection.

6. Add 450 µL of cell solution to each cuvette with 1-mL pipette tip. Gently pipette twice.

7. Set electroporator to 1,500 V and 25 F. Electroporate cells with two pulses with a 10-s interval. Expected time constant is 0.8–1 ms. Chill on ice for 2 min (see Note 4).

8. Using a sterile transfer pipette, transfer cells to a 25-cm² culture flask with 10-mL 29-13 medium prewarmed to 27°C and incubate with mild agitation at 27°C for 24 h.

3.2.3. Add Drug at 24-h Posttransfection

For single-plasmid transfection, add 10 μL of 2.5 mg/mL phleomycin (2.5 μg/mL final) or 10 μL of 10 mg/mL blasticidin (10 μg/mL final). For co-transfection, add both drugs. Incubate cells in 25 cm² culture flask with mild agitation at 27°C for 24 h.

3.2.4. Limiting Dilution

1. Add 100 μL of LD medium into wells 2–12 (all rows) of a 96-well flat-bottom plate.

2. Add 100 μL of LD medium into wells B1, D1, F1, and H1.

3. Add 300 μL of culture into wells A1, C1, E1, and G1.

4. Transfer 100 μL of culture from A1 to B1, C1 to D1, E1 to F1, and G1 to H1.

5. Using a multichannel pipetman, perform 1:1 serial dilutions (100 μL) from lane 1 through lane 12.

6. Starting from lane 12, add 100 μL of LD medium to each well to achieve 200 μL of total volume.

7. Incubate the 96-well plate at 27°C in humidified incubator with 5% CO_2.

8. Monitor growth under an inverted microscope. Expect to see cell growth in 10–14 days. If no live cells are present after 2 weeks, repeat the transfection.

9. When the well with highest dilution becomes confluent and the next dilution remains clear, transfer the entire 200-μL culture into 3 mL of LD medium in a 25-cm² flask and incubate upright without agitation for 24 h (see Note 5).

10. Start agitation (100 rpm) and continue until cell density reaches $5–10 \times 10^6$ cells/mL.

11. Add 3 mL of 29-13 medium with drugs and continue incubation for another 24 h.

12. Stabilize culture by diluting to 10^6 cells/mL two to fourfold every 24 h until division time is consistent.

3.3. Validation of iCODA/RNAi Cell Lines

Successful concurrent knockdown of endogenous protein and knock-in of exogenous protein can be verified at both the transcript and protein level. If antibodies are available, the depletion of endogenous protein and expression of longer TAP-tagged polypeptide can be verified by Western blotting (Fig. 2a). Parasite cells are collected by centrifugation at different induction time points, washed with phosphate buffered saline (PBS) buffer, and boiled in SDS loading buffer for 3 min. Cell extract is separated by SDS gel ($3–5 \times 10^6$ cells/well) and transferred by standard

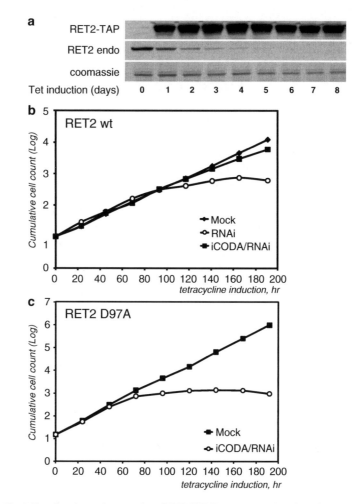

Fig. 2. Functional complementation of RET2 RNAi by co-expression of the RNAi-resistant transcript. (**a**) Western blotting analysis of RET2-iCODA/RNAi cells. (**b**) Growth kinetics of RET2-RNAi and RET2-iCODA/RNAi cell lines. (**c**) Growth kinetics of RET2-D97A-iCODA/RNAi cell line.

techniques. If antibodies are not accessible, Western blotting with commercially available PAP reagent (Sigma) which recognizes the Protein A moiety of the TAP tag and quantitative RT-PCR (qRT-PCR) can be used to confirm expression of TAP-tagged proteins and knockdown of the endogenous transcript, respectively. For total RNA isolation, 20–50 mL cultures are sufficient. Since transcripts derived from the iCODA-optimized sequence will be present in iCODA/RNAi cell lines, both qRT-PCR primers must selectively hybridize with the RNAi-targeted region of the wild-type transcript, but not with the iCODA-derived sequence. The knockdown of the endogenous transcript after ~48 h of RNAi should be normalized to mock-induced cells (see Note 6).

The RET2-RNAi-mediated growth phenotype is successfully rescued via co-expression of the iCODA-modified transcript (Fig. 2b). To confirm that the functional complementation requires RET2's enzymatic activity, a single amino acid mutation (D97A) has been introduced into the active site. As seen from Fig. 2c, the expression of the inactive RET2 does not counteract the RNAi growth inhibition phenotype.

3.4. Tandem Affinity Purification of Complexes from iCODA/RNAi Whole Cell Lysates

The following protocol is used to purify complexes from either iCODA/RNAi cells or from cells expressing the TAP-tagged protein. All purification steps are performed at 4°C.

1. Build up the 150 mL culture with cell density of ~10^7cells/mL (29-13 medium plus appropriate antibiotics). Avoid dilutions larger than five to tenfold as this can kill cells. Seed 2×850 mL cultures (1×10^6 cells/mL) in roller bottles (110×475 mm, Bellco) and add tetracycline to 1 µg/mL. Incubate in the roller apparatus until cell density reaches 12–15×10^6 cells/mL, typically 48–72 h.

2. Harvest cells in a fixed-angle rotor at $5,000 \times g$ for 10 min.

3. Resuspend pellet in 100 mL cold PBS pH 7.6 and repeat centrifugations.

4. Resuspend pellet in 50 mL of PBS, transfer into a conical centrifuge tube, collect cells at $3,000 \times g$ for 10 min, and proceed with complex purification. Pellet can be frozen in liquid nitrogen and stored at –80°C for later use.

5. Resuspend cells in 6-mL EB (final volume). Add NP-40 to 0.4% and 300 µL of Complete protease inhibitor (Roche).

6. Sonicate the extract three times at 9 W for 10 s.

7. Spin the extract at $100,000 \times g$ for 15 min at 4°C in a TLA 100 rotor and promptly transfer supernatant to a clean 15-mL conical tube.

8. Re-extract the pellet in 6 mL of EB (no NP-40) with sonication and centrifugation. Pool the two extracts – target volume is ~12 mL.

9. Prepare the IgG Sepharose resin. Transfer 0.3 mL of slurry (GE Healthcare) to a 15-mL tube and add 15-mL IBB. Centrifuge $300 \times g$ for 2 min (no brakes). Aspirate supernatant.

10. Filter the extract from step 8 through a syringe-driven low-protein binding membrane into the resin-containing tube from step 9. Incubate the extract with resin for 1 h with gentle agitation.

11. Transfer the material into a disposable column (Pierce). Reload three times to maximize recovery of resin. Collect flow-through.

12. Wash the column with 5–6 full volumes of IBB. Gently resuspend the resin with each wash and rinse the rims of the column well.

13. Wash IgG column with two column volumes of TCB.

14. Mix 150 units TEV protease, 20 μL Complete protease inhibitor, and 1.5 mL TCB. Close the column outlet, add TEV solution, seal both ends of the column with parafilm, and incubate overnight with gentle agitation.

15. Transfer 0.3 mL calmodulin resin into a 15-mL conical tube pretreated with 2% Tween 20 (see Note 7). Fill the tube with CBB, centrifuge at $300 \times g$ for 2 min and then aspirate the supernatant. Repeat wash once.

16. Drain the IgG column (~1.5 mL) directly onto the prewashed calmodulin resin.

17. Wash the IgG column with 3-mL CBB and collect the wash into the tube containing the calmodulin resin. Expect to recover ~4.5 mL total.

18. Add 6 μL 1 M $CaCl_2$ to the suspension and incubate the tube with gentle agitation for 1 h.

19. Transfer the suspension to a disposable column pretreated with 2% Tween 20. Reload flow through 3–5 times. Collect the flow-through.

20. Wash the column with 5–6 full column volumes with CBB with resin.

21. Close the bottom of the column. Apply 0.5 mL CEB to the resin and incubate for 10 min at room temperature with periodic gentle mixing to resuspend the resin. Collect the elution in a low-protein binding microcentrifuge tube. Repeat thrice (see Note 8).

22. Flash-freeze fractions in liquid nitrogen and store at −80°C.

23. To visualize the complexes, load ~10 μL of eluted fractions onto an SDS-PAGE gel and stain the gel with Sypro Ruby.

Protein profiles of RECCs purified from RET2-mhTAP, RET2-iCODA-TAP, and RET2-iCODA/RNAi cells are indistinguishable (Fig. 3). Collect aliquots at all purification steps starting from total cell lysate. Quantitative Western blotting with anti-CBP antibody (GenScript) can be used to estimate purification yield for the bait protein, which is expected to be 25–30%. For mass spectrometry analysis, final fraction is precipitated by adding trichloric acid to 20% and deoxycholate to 0.1%, washed with acetone twice and digested with trypsin by standard protocols.

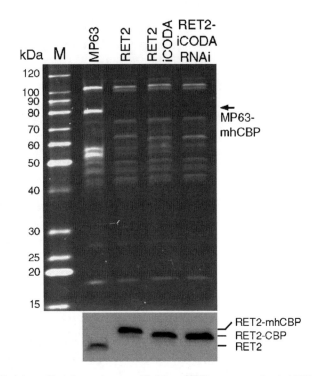

Fig. 3. Protein profiles of complexes purified from RET2 overexpression (mhTAP), RET2-iCODA-TAP, and RET2-iCODA/RNAi cell lines. (*Top panel*) TAP-purified complexes were separated on an 8–16% gradient SDS-PAGE gel and stained with Sypro Ruby staining. Cell lines are listed above the gel. mhCBP – 6His plus calmodulin binding peptide which remain on the tagged protein upon TEV protease release from the IgG resin. (*Bottom panel*) TAP-purified complexes were analyzed by Western blotting with anti-RET2 antibodies. Affinity purification of structural RECC subunit (MP63) was used to demonstrate gel mobility of the endogenous RET2.

4. Notes

1. Many commercial batches of FBS are contaminated with tetracycline. Some manufacturers (BD, Invitrogen) claim to have a tetracycline-free serum which does not always guarantee the leak-proof transcriptional control of RNAi expression. Apparently, mild to severe growth inhibition phenotypes can be induced in *T. brucei* RNAi cells by tetracycline present at levels undetectable with current testing methods. It is advisable to collect several serum samples and test for growth kinetics of an RNAi cell line in which an essential gene is targeted (available from the author's laboratory). Typically, serum supporting the fastest growth would be most suited for cloning procedures and maintenance of uninduced cells.

2. When preparing TCB, DTT should be added immediately before applying to the IgG beads.

3. Co-transfection of both TAP and RNAi constructs requires less time and is a default protocol. If obtaining clones resistant to all four drugs is unsuccessful, sequential transfection is recommended. First, transfect 29-13 cells with the protein expression construct (blasticidin resistance) and stabilize cells in a 10 mL of culture without clonal selection (~3 weeks). Second, transfect cells with the RNAi construct and proceed with clonal selection.

4. Incubation of the electroporated cells on ice longer than 5 min is not recommended. When performing multiple transfections, electroporations should be staggered to decrease the time between electroporation and inoculation into medium. If an electric arc occurs, expect to obtain fewer or no clones.

5. The medium color changes with cell growth from red at low density to yellow in the stationary culture. Cells can be ready for passage at 3–4 weeks posttransfection. Care should be taken to avoid oversaturated (yellow) cultures. Transfer at least 3–5 clones, but continue incubating the 96-well plate for a few more days to ensure lack of cell growth in wells with higher dilutions. If this occurs, discard previously collected clones and transfer new ones from the highest dilution wells.

6. For RNAi knockdown of essential genes, induced and mock-induced cell should be harvested at the same cell density. Typically, RNAi is induced at 10^6 cells/mL and the culture is maintained for 24–48 h; cell density should not exceed $5–6 \times 10^6$ cells/mL.

7. Pretreatment of plastic tubes and columns during the calmodulin binding step with 2% Tween 20 is recommended to improve purification yields. Pretreated plastic should be rinsed with large amounts of distilled water before use.

8. Fractions 1 and 2 should contain most of the eluted material. The first 50–100 µL of the Fraction 1 is the void volume and does not need to be collected.

Acknowledgments

We thank George Cross and Elisabetta Ullu for kind gifts of cell lines and plasmids. This work was supported by the NIH grants RO1AI064653 to RA and R01CA112560 to RHL.

References

1. Aphasizhev, R., Aphasizheva, I., Nelson, R. E., Gao, G., Simpson, A. M., Kang, X., Falick, A. M., Sbicego, S., and Simpson, L. (2003) Isolation of a U-insertion/deletion editing complex from *Leishmania tarentolae* mitochondria. *EMBO J.* **22**, 913–924.

2. Panigrahi, A. K., Schnaufer, A., Ernst, N. L., Wang, B., Carmean, N., Salavati, R., and Stuart, K. (2003) Identification of novel components of *Trypanosoma brucei* editosomes. *RNA* **9**, 484–492.

3. Panigrahi, A. K., Ernst, N. L., Domingo, G. J., Fleck, M., Salavati, R., and Stuart, K. D. (2006) Compositionally and functionally distinct editosomes in *Trypanosoma brucei*. *RNA* **12**, 1038–1049.

4. Weng, J., Aphasizheva, I., Etheridge, R. D., Huang, L., Wang, X., Falick, A. M., and Aphasizhev, R. (2008) Guide RNA-binding complex from mitochondria of Trypanosomatids. *Mol. Cell* **32**, 198–209.

5. Puig, O., Caspary, F., Rigaut, G., Rutz, B., Bouveret, E., Bragado-Nilsson, E., Wilm, M., and Seraphin, B. (2001) The tandem affinity purification (TAP) method: a general procedure of protein complex purification. *Methods* **24**, 218–229.

6. Aphasizhev, R., Aphasizheva, I., Nelson, R. E., and Simpson, L. (2003) A 100-kD complex of two RNA-binding proteins from mitochondria of *Leishmania tarentolae* catalyzes RNA annealing and interacts with several RNA editing components. *RNA* **9**, 62–76.

7. Schnaufer, A., Panigrahi, A. K., Panicucci, B., Igo, R. P., Salavati, R., and Stuart, K. (2001) An RNA ligase essential for RNA editing and survival of the bloodstream form of *Trypanosoma brucei*. *Science* **291**, 2159–2161.

8. Trotter, J. R., Ernst, N. L., Carnes, J., Panicucci, B., and Stuart, K. (2005) A deletion site editing endonuclease in *Trypanosoma brucei*. *Mol. Cell* **20**, 403–412.

9. Carnes, J., Trotter, J. R., Ernst, N. L., Steinberg, A., and Stuart, K. (2005) An essential RNase III insertion editing endonuclease in *Trypanosoma brucei*. *Proc. Natl. Acad. Sci. U. S. A.* **102**, 16614–16619.

10. Ngo, H., Tschudi, C., Gull, K., and Ullu, E. (1998) Double-stranded RNA induces mRNA degradation in *Trypanosoma brucei*. *Proc. Natl. Acad. Sci U. S. A.* **95**, 14687–14692.

11. Gossen, M. and Bujard, H. (1992) Tight control of gene expression in mammalian cells by tetracycline-responsive promoters. *Proc. Natl. Acad. Sci. U. S. A.* **89**, 5547–5551.

12. Wirtz, E., Leal, S., Ochatt, C., and Cross, G. A. (1999) A tightly regulated inducible expression system for conditional gene knock-outs and dominant-negative genetics in *Trypanosoma brucei*. *Mol. Biochem. Parasitol.* **99**, 89–101.

13. Salavati, R., Ernst, N. L., O'Rear, J., Gilliam, T., Tarun, S. Jr., and Stuart, K. (2006) KREPA4, an RNA binding protein essential for editosome integrity and survival of *Trypanosoma brucei*. *RNA* **12**, 819–831.

14. Aphasizheva, I., Ringpis, G. E., Weng, J., Gershon, P. D., Lathrop, R. H., and Aphasizhev, R. (2009) Novel TUTase associates with an editosome-like complex in mitochondria of *Trypanosoma brucei*. *RNA* **15**, 1322–1337.

15. Rusconi, F., Durand-Dubief, M., and Bastin, P. (2005) Functional complementation of RNA interference mutants in trypanosomes. *BMC Biotechnol.* **5**, 6.

16. Deng, J., Schnaufer, A., Salavati, R., Stuart, K. D., and Hol, W. G. (2004) High resolution crystal structure of a key editosome enzyme from *Trypanosoma brucei*: RNA editing ligase 1. *J. Mol. Biol.* **343**, 601–613.

17. Deng, J., Ernst, N. L., Turley, S., Stuart, K. D., and Hol, W. G. (2005) Structural basis for UTP specificity of RNA editing TUTases from *Trypanosoma brucei*. *EMBO J.* **24**, 4007–4017.

18. Larsen, L. S., Wassman, C. D., Hatfield, G. W., and Lathrop, R. H. (2008) Computationally optimised DNA assembly of synthetic genes. *Int. J. Bioinform. Res. Appl.* **4**, 324–336.

19. Aphasizhev, R., Aphasizheva, I., and Simpson, L. (2003) A tale of two TUTases. *Proc. Natl. Acad. Sci. U. S. A.* **100**, 10617–10622.

20. Ernst, N. L., Panicucci, B., Igo, R. P., Jr., Panigrahi, A. K., Salavati, R., and Stuart, K. (2003) TbMP57 is a 3′ terminal uridylyl transferase (TUTase) of the *Trypanosoma brucei* editosome. *Mol. Cell* **11**, 1525–1536.

21. Wickstead, B., Ersfeld, K., and Gull, K. (2002) Targeting of a tetracycline-inducible expression system to the transcriptionally silent minichromosomes of *Trypanosoma brucei*. *Mol. Biochem. Parasitol.* **125**, 211–216.

22. Redmond, S., Vadivelu, J., and Field, M. C. (2003) RNAit: an automated web-based tool for the selection of RNAi targets in *Trypanosoma brucei*. *Mol. Biochem. Parasitol.* **128**, 115–118.

23. Kelly, S., Reed, J., Kramer, S., Ellis, L., Webb, H., Sunter, J., Salje, J., Marinsek, N., Gull, K., Wickstead, B., and Carrington, M. (2007) Functional genomics in *Trypanosoma brucei*:

a collection of vectors for the expression of tagged proteins from endogenous and ectopic gene loci. *Mol. Biochem. Parasitol.* **154**, 103–109.

24. Jensen, B. C., Kifer, C. T., Brekken, D. L., Randall, A. C., Wang, Q., Drees, B. L., and Parsons, M. (2007) Characterization of protein kinase CK2 from *Trypanosoma brucei. Mol. Biochem. Parasitol.* **151**, 28–40.

25. Wassman, C. D., Tam, P. Y., Lathrop, R. H., and Weiss, G. A. (2004) Predicting oligonucleotide-directed mutagenesis failures in protein engineering. *Nucleic Acids Res.* **32**, 6407–6413.

26. Sharp, P. M. and Li, W. H. (1986) Codon usage in regulatory genes in *Escherichia coli* does not reflect selection for 'rare' codons. *Nucleic Acids Res.* **14**, 7737–7749.

27. Gutman, G. A. and Hatfield, G. W. (1989) Nonrandom utilization of codon pairs in *Escherichia coli. Proc. Natl. Acad. Sci. U. S. A.* **86**, 3699–3703.

28. Hatfield, G. W. and Roth, D. A. (2007) Optimizing scaleup yield for protein production: computationally optimized DNA assembly (CODA) and translation engineering. *Biotechnol. Annu. Rev.* **13**, 27–42.

29. Irwin, B., Heck, J. D., and Hatfield, G. W. (1995) Codon pair utilization biases influence translational elongation step times. *J. Biol. Chem.* **270**, 22801–22806.

30. Mathews, D. H., Sabina, J., Zuker, M., and Turner, D. H. (1999) Expanded sequence dependence of thermodynamic parameters improves prediction of RNA secondary structure. *J. Mol. Biol.* **288**, 911–940.

Part II

Adenosine to Inosine RNA Editing

Chapter 3

Perturbing A-to-I RNA Editing Using Genetics and Homologous Recombination

Cynthia J. Staber, Selena Gell, James E.C. Jepson, and Robert A. Reenan

Abstract

Evidence for the chemical conversion of adenosine-to-inosine (A-to-I) in messenger RNA (mRNA) has been detected in numerous metazoans, especially those "most successful" phyla: Arthropoda, Mollusca, and Chordata. The requisite enzymes for A-to-I editing, ADARs (adenosine deaminases acting on RNA) are highly conserved and are present in every higher metazoan genome sequenced to date. The fruit fly, *Drosophila melanogaster*, represents an ideal model organism for studying A-to-I editing, both in terms of fundamental biochemistry and in relation to determining adaptive downstream effects on physiology and behavior. The *Drosophila* genome contains a single structural gene for ADAR (*dAdar*), yet the fruit fly transcriptome has the widest range of conserved and validated ADAR targets in coding mRNAs of any known organism. In addition, many of the genes targeted by *dADAR* have been genetically identified as playing a role in nervous system function, providing a rich source of material to investigate the biological relevance of this intriguing process. Here, we discuss how recent advances in the use of ends-out homologous recombination (HR) in *Drosophila* make possible both the precise control of the editing status for defined adenosine residues and the engineering of flies with globally altered RNA editing of the fly transcriptome. These new approaches promise to significantly improve our understanding of how mRNA modification contributes to insect physiology and ethology.

Key words: *Drosophila melanogaster*, RNA editing, Homologous recombination

1. Introduction

1.1. A-to-I Editing in Drosophila melanogaster

A-to-I RNA editing has been shown to be crucial for normal behavior and/or development in a range of model genetic organisms (1–4). The deamination of adenosine by ADARs requires the formation of a double-strand RNA intermediate, which acts as the substrate for an ADAR dimer (5, 6). In *Drosophila*, a wide range of adenosines in the coding regions of mRNAs undergo

Ruslan Aphasizhev (ed.), *RNA and DNA Editing: Methods and Protocols*, Methods in Molecular Biology, vol. 718,
DOI 10.1007/978-1-61779-018-8_3, © Springer Science+Business Media, LLC 2011

A-to-I editing. Because inosine is recognized by the cellular machinery as guanosine (7), editing can lead to amino-acid coding changes, and does so at a majority of known editing sites in flies. Thus, a spectrum of proteins that are not directly encoded by the genome are created through a simple chemical modification. As mentioned earlier, the binding of dADAR to coding mRNAs is facilitated by the formation of a dsRNA structure comprising the region surrounding the edited exon and a *cis*-acting complementary sequence, the ECS (exon complementary sequence), which is generally located in an adjacent intron (8, 9). Intriguingly, targets of dADAR are highly enriched for transcripts whose primary roles are nervous system function, including a range of voltage- and ligand-gated ion channels and several regulators of neurotransmitter release from the presynaptic terminal (10–13). Correspondingly, loss of dADAR activity results in profound behavioral defects, including extreme uncoordination, seizures, temperature-sensitive paralysis, and progressive adult-stage neuro-degeneration (3).

A significant constraint on research into A-to-I editing in insects has been the difficulty in using traditional transgenic expression techniques to unravel how individual editing sites contribute to the functioning of the *Drosophila* nervous system. There are two reasons that conventional transgenic reverse genetics fails to provide traction on this crucial issue. Firstly, the extreme phenotype of *dAdar* null flies (which lack editing at all sites) precludes analysis of complex behaviors. Secondly, although it is technically feasible to generate flies that are null for an edited mRNA of interest, and subsequently express unedited or edited cDNA transgenes in that background, this approach still removes all endogenous regulation of the mRNA in question, making interpretation of behavioral data difficult. Since editing is, in essence, a polygenic process whose output may affect individual genes in subtle ways, what is desired is a method that introduces minimal perturbations in other aspects of gene expression (e.g., transcript levels, alternative splicing). In addition, although several studies have modeled or experimentally determined the structures that direct editing either in silico or *in vitro* (9, 14–16), no verification of these predictions has been performed in the true biological context of RNA editing: the whole organism.

The ability to selectively engineer genetic loci by HR *in vivo* solves both of the above issues, since endogenous regulation is maintained. Further, the physiological consequences of altering editing at specific sites in an otherwise wild-type genetic background can be investigated at the molecular, cellular, and whole-organism levels. Here, we present a detailed methodology for targeted ends-out homologous recombination (HR) (17, 18) that facilitates such alterations. Before doing so, we briefly outline some of the broad experimental strategies that researchers may choose to use to generate specifically altered genetic backgrounds to study the biological consequences and evolutionary implications of RNA editing.

1.2. Deletion of Editing at Specific Adenosines

We envision at least four separate methods that could be used to abolish editing at an adenosine of choice. Specifically edited codons may be mutated to synonymous codons lacking the edited adenosine while conserving the amino-acid sequence. However, this strategy only works for five codons. Four codons contain first position A residues that can be mutated to synonymous codons encoding the same amino acid but lacking the edited adenosine, serine (e.g., **A**GT/C → **T**CT/C), and arginine (e.g., **A**GA/G → **C**GA/G). The remaining codon for which this method can work is the third position edit for **I** → **M** where the AT**A** codon can be mutated to AT(**T/C**) to render editing at the third position an impossibility.

Another minimally invasive method that could abolish or severely restrict editing involves taking advantage of ADAR enzyme 5′ neighbor preferences. Numerous studies have revealed an extreme bias against guanosine (G) as a 5′ neighbor to edited adenosines. Therefore, when feasible, one could alter editing of A residues in first positions of codons by changing the third position of the previous codon to a G.

In cases where the codon itself cannot be mutated, alternative strategies must be employed. A prerequisite for these methods are robust predictions as to the precise *cis*-acting elements that direct editing. Fairly accurate predictions can be initially gained through the combined use of comparative genomics and computational software such as Mfold (http://mfold.burnet.edu.au/) or Sfold (http://sfold.wadsworth.org). If the adenosine of interest is edited in other *Drosophilid* species, the elements that direct editing are also likely to be conserved. Several studies have used this strategy to define *cis*-acting elements by simply aligning introns that flank the edited exon in orthologous genes from multiple *Drosophilid* or insect genome sequences, and identifying highly conserved sequence stretches (8, 9, 16). Once such sequences are identified, complementarity to the sequence surrounding the edited adenosine can be tested in silico using MFold or Sfold, which predict the secondary structure of the RNA (it should be noted that while both software programs are useful for prediction of simple fold-back dsRNA structures, they cannot accurately predict more complex structures such as pseudo knots or higher order junctions) (9). Nevertheless, when in silico predictions of structure using significant regions of mostly nonconserved sequence base pair in the highly conserved noncoding elements with regions around the editing site, some degree of confidence can be placed in the prediction.

After the location of the hypothesized *cis*-acting element is determined, two approaches can be used. Firstly, deletion of the entire *cis*-element should act to completely abolish editing at the target adenosine. Secondly, one can modify residues that base pair with the edited adenosine. Although ADARs have only minimal substrate specificity in terms of the sequence surrounding the edited adenosine, the opposing residue to the adenosine is of significant catalytic importance. Adenosines in A-U base pairs and A-C mis-matches are efficient substrates for ADARs, while ADARs are inhibited by mutation of A-U/A-C to A-A/A-G mis-matches (19). Therefore, the U or C residue predicted to lie opposite the edited adenosine could be mutated to a purine residue. This approach is inherently more risky than deleting the entire *cis*-element, as it requires in silico structural predictions to be extremely accurate. However, it avoids the deletion of sequence stretches within an intron, which may potentially have unintended consequences for alternative splicing and gene expression. Another approach is to introduce selected mutations into the ECS intronic element that substantially un-pair the predicted duplex without changing the spacing of any elements (9). This method also was shown to allow the modeling of structurally predicted changes in species-specific editing structures to allow for structure-guided introduction of new editing sites.

It is worth mentioning that in a minority of cases, an exonic sequence alone can fold into an ADAR substrate (e.g., in *dAdar*'s own transcript (20)). In these cases, synonymous mutations could be made in the exon itself that disrupt base pairing to the region surrounding the edited adenosine while maintaining the correct amino acid sequence.

1.3. "Hard-Wiring" Editing at Defined Adenosines

Editing at many sites is tightly controlled both temporally and spatially (8, 16, 21, 22), but the physiological reasons for such stringent regulation are unclear. Using HR, adenosines of choice may be converted to guanosines, thus effectively "hard-wiring" the fully edited state into the genome and removing all developmental and spatial regulation. The consequences of such deregulation can then be examined at the organismal level with the appropriate genetic controls.

1.4. Testing Structural Predictions

If deletion of a hypothesized *cis*-element (as described earlier) leads to the loss of detectable editing at an adenosine, this provides direct evidence as to the sequences that direct editing in vivo. However, it does not define the precise structure that acts as an ADAR substrate. In concert with in silico predictions, HR can be used to directly test structural hypotheses. For example, short sequence stretches in *cis*-elements predicted to bind closely to the edited adenosines can be replaced with restriction endonuclease (RE) sites, thus disrupting dsRNA formation. Complementary alterations can then be made in the coding sequence that restore base pairing to the RE site. This mutation/counter-mutation approach has been successfully used *in vitro* to prove structural predictions (9, 15). HR allows similar approaches to be performed *in vivo* and will undoubtedly be crucial for assessing effects of RNA structural alterations where the proposed RNA structures are complex and dynamic.

1.5. Epitope Tagging and Engineered Editing Reporter Systems

HR can also be used to introduce small epitope or affinity tags (e.g., hemagglutinin antigen-HA, FLAG, or 6HIS) into ADAR or an edited target of ADAR. Such methods could allow for the localization of ADAR activity *in vivo* or the potential effects of ADAR activity on ADAR target gene product localization. For instance, since editing has been shown in mammalian systems to effect the trafficking of GluR-B subunits through the endoplasmic reticulum, tagging an ADAR target gene could demonstrate an ADAR dependence on protein localization.

Another approach might involve utilizing a read-through approach. Since ADAR activity could lead to the programmed read-through of a TAG stop codon (reassigning it to TGG or tryptophan), one could design a fusion protein with, for instance, green fluorescent protein, downstream of the edited TAG such that ADAR activity (conversion of TAG to TGG) results in a fluorescent signal.

2. Materials

2.1. Fly Stocks for Mapping and HR

1. The following fly stocks used in the HR procedure can be obtained from the Bloomington Stock Center (www.flystocks. bio.indiana.edu/). Stock center numbers and complete genotypes are given after the abbreviated stock name.

y w;FLP-I-SceI, noc^{Sco}/CyO (BSC 6934)

y^1 *w*; P{ry$^{+t7.2}$=70FLP}11 P{ v$^{+t1.8}$=70I -SceI}2B noc^{Sco}/CyO, S^2*

y w;FLP-I-SceI/TM6 (BSC 6935)

y^1 *w*; P{ry$^{+t7.2}$=70FLP}23 P{v$^{+t1.8}$=70I-SceI}4A/TM6*

y w ey-FLP (BSC 5580)

y^{d2} *w^{1118} P{ry[+t7.2]=ey-FLP.N}2*

y w Cre; noc^{Sco}/CyO (BSC 766)

y^1 *w^{67c23} P{y$^{+mDint2}$=Crey}1b; noc^{Sco}/CyO*

y w Cre;D*/TM3, Sb (BSC 851)

y^1 *w^{67c23} P{y$^{+mDint2}$=Crey}1b; D*/TM3, Sb^1*

y w; noc^{Sco}/CyO, Cre (BSC 1092)

y^1 *w^{67c23}; noc^{Sco}/CyO, P{w^{+mC}=Crew}DH1*

w^1 (BSC 145)

w^{1118} (BSC 3605)

A number of balancer chromosomes used for mapping and stocking purposes are also available from the BSC. Our lab uses FM7c (X), FM7a (X), w;CyO/Sco (second), and w;TM3/TM6 (third), but any appropriately marked balancers may be used. All stocks are maintained at room temperature (~22°C).

2. Standard cornmeal/molasses fly food in both vials and bottles are used for stock maintenance and targeting crosses. The BSC website has several detailed fly food recipes and cooking instructions. We use a cornmeal/molasses/brewer's yeast recipe and add propionic acid (10 ml/3 l) to combat bacterial contamination and tegosept (11 mls 20% in 3l) to combat mold. Fresh food (1–3 days old) is best for the targeting and mapping crosses. Food up to two weeks old is acceptable for stock transfers provided it has been stored at 4°C and has not dried out.

2.2. Cloning and Site-Directed Mutagenesis of Arms

1. High-fidelity Taq DNA polymerase such as Platinum Taq HF (Inivitrogen) or Phusion Taq (New England Biolabs) (see Note 1).

2. Topo-TA or Topo-Blunt Cloning Kit (Invitrogen).

3. Quik-Change II XL Site-directed Mutagenesis Kit (Stratagene).

4. LB-Amp media, plates, and broth as follows: for 1 l, dissolve 10 g tryptone, 5 g yeast extract, and 10 g NaCl in 950 ml deionized water. Adjust pH to 7.0 with NaOH and bring volume to 1 l with water. Autoclave on a liquid cycle for 20 min. Allow to cool to <55°C before adding ampicillin at 50 μg/ml. For plates, add 15 g agar per liter before autoclaving.

5. Wizard Gel and PCR Cleanup Kit (Promega).

6. Pure Yield Plasmid Mini- and Midi-prep Kits (Promega).

7. T4 DNA Ligase, I Kb ladder, and restriction enzymes: AscI, BsiWI, Acc65I, Not I, and I-SceI (New England Biolabs).

8. Amplification primers for arms 1 and 2 designed to incorporate restriction enzyme cloning sites (Fig. 1) and mutagenic primers designed according to QuikChange Manual (Stratagene).

9. P[w25.2] vector (Drosophila Genomics Resource Center).

10. Maxwell 16 and Maxwell 16 Tissue DNA Purification kits (Promega) or other genomic DNA prep kits.

11. RNAse A, 20 mg/ml. Dilute to 20 ug/ml in water and use in place of elution buffer from Maxwell kit.

2.3. PCR

1. Standard Taq (New England Biolabs) and Go-Taq HS (Promega) (see Note 1)

2. Validation primers (2–3 recommended) designed to genomic locus flanking arms 1 and 2.

a PCR amplify each arm with primers that add the necessary restriction sites

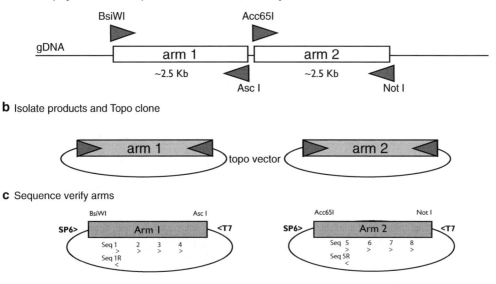

b Isolate products and Topo clone

c Sequence verify arms

d Excise arms from Topo clones using engineered restrictions sites and ligate sequentially into P[w25.2] vector

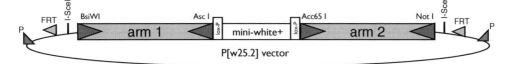

Fig. 1. Overview of homology arm amplification and cloning steps. (**a**) PCR amplifie each arm from wild-type genomic DNA using primers that add the necessary restriction sites at either end (shaded triangles). (**b**) Isolate PCR products and Topo clone. (**c**) Sequence verify arm inserts. (**d**) Excise arms from Topo clones and sequentially ligate into prepared P[w25.2] vector.

2.4. Sequencing

1. Big Dye Terminator v. 3.1 (Applied Biosystems). Aliquot enzyme and buffer into small volumes (~200 μl) to avoid multiple freeze/thaw cycles. Store at –20°C.

2. Agencourt Clean Seq (Agencourt Bioscience Co.). Store at 4°C and vortex to resuspend beads before each use.

3. 80% ethanol (prepared fresh daily).

4. Agencourt SPRIPlate® 96R – Ring Magnet Plate (Agencourt Bioscience Co.).

5. Topo vector primers (5'–3'):

   TOPO-SP6 GCCAAGCTATTTAGGTGACACTATAG

   TOPO-T7 GAATTGTAATACGACTCACTATAGGG

6. P[w25.2] vector primers (5'–3'):

   pW-Not 1 CACTGTTCACGTCGCACTCGAGGGTAC

   pW-Not 2 GCACTCGAGAGCTCGTTACAGTCCG

   pW-Bsi 1 CGCACCGGACTGTAACGAGCTAC

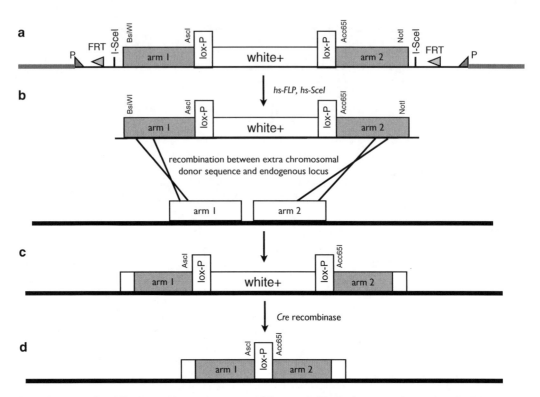

Fig. 2. Summary of mobilization and integration events. (**a**) Transgenic P[w25.2] construct inserted randomly in genome is mobilized by *hs-FLP* and linearized by hs-*Sce-I*. (**b**) The extrachromosomal donor cassette recombines at the endogenous locus via the homology arms. (**c**) Targeted locus pre-*Cre*. *Cre* recombinase excises the sequence between the lox-P sites leaving behind one lox-P site and the flanking six-frame stop sequence. (**d**) Targeted locus post-*Cre* with mini-white gene removed. Engineered arms are represented by shaded boxes and endogenous arms by white boxes (**a–d**). Six-frame stop sequence not shown.

pW-Bsi 2 GGCGACTCAACGCAGATGCCGTACC

pW-Asc 1 GTATGCTATACGAAGTTATCTAGACTAGTC
TAGGGCG

pW-Asc 2 GCTTGGCTGCAGGTCGACTCTAGAGG

pW-Asc 3 CGATCATTCATTATTCGCTGCATGAATTAGC

pW-Acc 1   CATTATACGAAGTTATCTAGACTAGTCTA
GGGTAC

pW-Acc 2 GACGCTCCGTCGACGAAGCGCCTC

pW-Acc 3 GCTCAGCTTGCTTCGCGATGTGTTCAC

7. Gene-specific sequencing primers to cover arms 1 and 2 (Fig. 5).

2.5. Initiating HR via Heat Shock

1. Narrow diameter polystyrene food vials (Applied Scientific or Genesee Scientific) with food poured at a uniform height of ~3 cm.

2. Nalgene unwire racks for 30 mM tubes (Thermo Scientific Nalgene 5970-0230, 5972-0030).

3. Circulating water bath that will accommodate Nalgene racks set to 38–39°C.

3. Methods

Ends-out HR allows surgical insertion, deletion, or alteration of specific sequences within an endogenous locus, making it an essential new tool for *Drosophila* geneticists. In the case of RNA editing, it allows hardwiring of specific edited adenosines or analysis of mutations and counter-mutations in ECS regions and other proposed regulatory sequences. Ends-out HR permits a nearly unbiased analysis of the effect of mutations *in vivo* without the requirement of controlling for the unintended side effects that frequently plague even the most carefully conducted transgenic studies, including position effect, expression levels, or background due to expression of the endogenous wild-type gene.

HR begins with construction of a targeting vector containing two homology arms specific to the gene of interest. The random integration of this targeting vector in the fly genome followed by the subsequent excision and targeted re-integration at the desired endogenous locus completes the first phase of HR. Figure 1 shows the molecular steps involved in constructing the P[w25.2] transgenic vectors. Figure 2 shows the events that occur during the various fly crosses required to target the P[w25.2] construct to the endogenous locus. Both figures are referred to throughout this chapter. Figure 3 shows an approximate timeline of the HR crosses from initiating excision through validating potential integration events.

Vector construction begins by amplifying two homology arms (~2.5 Kb each) from genomic DNA that are specific to the gene of interest and cloning them into the Topo vector (Fig. 1a, b). Once sequence verified (Fig. 1c), these arms serve as a template for site-directed mutagenesis to engineer specific mutations in one or both arms. The arms, both wild type and mutant, will be sub-cloned into the P[w25.2] vector (Fig. 1d) to create separate wild-type and mutant constructs. The P[w25.2] constructs are inserted randomly into the genome using standard fly transformation techniques (23). Once mapped and stocked, transgenic lines are targeted by crossing to stocks that express *FLP* and *I-SceI* which excise the homology arm region creating an extrachromosomal recombinogenic donor cassette (Fig. 2a, b). In addition to the homology arms, this extrachromosomal donor cassette carries the mini-white eye color marker flanked by lox-P sites.

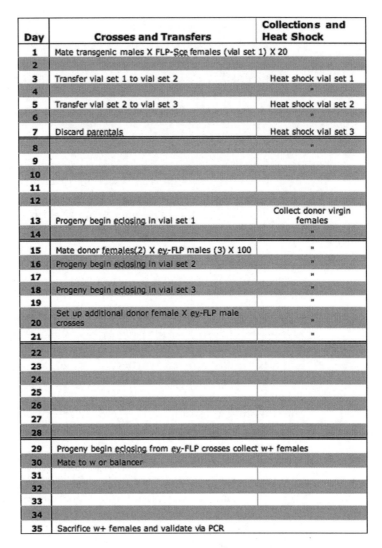

Day	Crosses and Transfers	Collections and Heat Shock
1	Mate transgenic males X FLP-Sce females (vial set 1) X 20	
2		
3	Transfer vial set 1 to vial set 2	Heat shock vial set 1
4		"
5	Transfer vial set 2 to vial set 3	Heat shock vial set 2
6		"
7	Discard parentals	Heat shock vial set 3
8		"
9		
10		
11		
12		
13	Progeny begin eclosing in vial set 1	Collect donor virgin females
14		"
15	Mate donor females(2) X ey-FLP males (3) X 100	"
16	Progeny begin eclosing in vial set 2	"
17		"
18	Progeny begin eclosing in vial set 3	"
19		"
20	Set up additional donor female X ey-FLP male crosses	"
21		"
22		
23		
24		
25		
26		
27		
28		
29	Progeny begin eclosing from ey-FLP crosses collect w+ females	
30	Mate to w or balancer	
31		
32		
33		
34		
35	Sacrifice w+ females and validate via PCR	

Fig. 3. Timeline of HR fly crosses. Timeline shows approximates dates of crosses and collections beginning with the mating of donor transgenic males and *FLP-Sce* females through sacrificing targeted flies for validation PCR. Timing assumes stocks are maintained at ~22°C.

Recombination occurs between the homology arms and the endogenous arm region at the target locus, inserting the wild type or mutant sequence and the intervening vector sequence (Fig. 2b, c).

The second phase of HR begins with validation of targeting events. Proper integration of the construct is confirmed by PCR, sequencing, and/or restriction enzyme digests. The mini-white sequence can then be removed by crossing the flies to a line expressing Cre recombinase (Fig. 2c, d). At the completion of HR, the target locus will contain the mutations of interest as well as 76 bp of vector sequence that replaces ~76 bp of endogenous noncoding sequence (Fig. 2d). Wild-type controls are done in

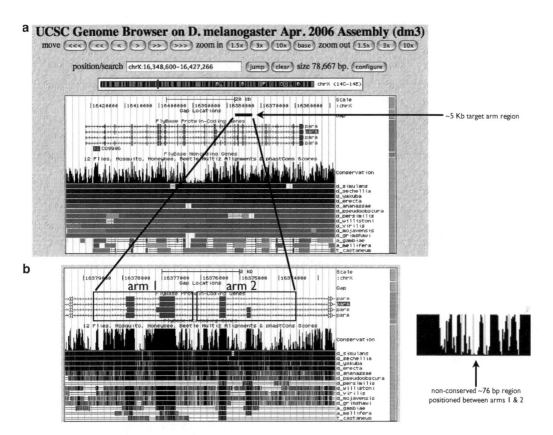

Fig. 4. Designing homology arms using the UCSC Genome Browser. (a) The *para* locus, spanning ~68 Kb, encodes a voltage-gated sodium channel containing a number of edited adenosines. A 5-Kb region spanning several editing sites is chosen. (b) Zooming in on the region of interest and centering arms across a small region of nonhomology (enlarged at *right*).

each case to confirm that this small sequence replacement/ insertion has no effect on expression or function of the endogenous gene.

Although many techniques such as PCR, cloning, site-directed mutagenesis, and sequencing are utilized in making the constructs and generating the transgenic flies, these are standard molecular biology techniques not modified in any particular way for HR. As this chapter is intended to focus on HR, here we describe in detail the methods critical to a successful HR, specifically construct design and the actual targeting and analysis, and only briefly detail the other techniques.

3.1. Design and Amplification of Targeting Arms

The UCSC Genome browser (http://genome.ucsc.edu/) is useful for designing arm regions as it shows the degree of conservation at the nucleotide level across multiple *Drosophila* and other insect species (Fig. 4). Begin by identifying approximately 5 Kb of genomic sequence (2.5 Kb/arm) spanning the region to be mutated. Center the sequence across a nonconserved dispensable region such as an intron or intergenic region that can be replaced by 76 bp of vector sequence without likely affecting gene function. The nonconserved region need not be exactly 76 bases but should be relatively close in size to minimize potential disruption of the gene posttargeting. While we practice replacement of 76 bases of nonconserved sequence to preserve spacing, this is obviously an experimenter's preference. Nonconserved sequence can function to spatially separate coding sequence or regulatory regions, thus disruption of this distance may affect gene expression. If possible, center the arms such that prospective mutations lie close to the center of the homology arms as this increases the likelihood of their inclusion in the integration event.

Another benefit of the UCSC browser is the ability to extract sequence with user-defined parameters, such as choosing the length of upstream or downstream sequence to include or having exons display in upper case and introns in lower case. The UCSC Bioinformatics Group sponsors a number of free tutorials that are helpful in learning to use the UCSC browser (http://www.openhelix.com/downloads/ucsc/ucsc_home.shtml).

When designing the arms, several factors must be kept in mind regarding the eventual insertion of the arms into the P[w25.2] transformation vector (Fig. 5). Arms may be cloned in either orientation within the P[w25.2] vector, but the 5′–3′ orientation of both arms with respect to the endogenous locus must be maintained. Each arm sequence must also be free of the restriction enzyme sites that will eventually be used to insert the arm in the P[w25.2] vector. For example, when working in the "+ orientation" (Fig. 5a), BsiWI and AscI restriction sites flank arm 1 and will be used to cut the arm from the Topo vector and insert it into the P[w25.2] vector. Thus, neither of those sites can lie within the arm sequence. The same holds true for arm 2, which may not contain an Acc65I or Not I site. If either arm carries an internal site associated with the cloning sites for the opposite arm, attention must be paid to the order in which the arms are sub-cloned into P[w25.2]. If arm 1 contains a Not I or Acc65 I site then arm 2 must be cloned into P[w25.2] first and arm 1 cloned in last. Conversely, if arm 2 contains an Asc I or BsiWI site, then arm 1 must be cloned first (see Note 2).

1. Prepare genomic DNA from a wild-type *Drosophila* stock to use as a template for amplification of arm sequence. The Maxwell 16 Tissue DNA Kit is used with the Maxwell 16 robot to isolate genomic DNA from *Drosophila* tissue (24). Instead of the elution buffer provided with the Maxwell kit, we elute in water containing 20 µg/ml of RNAse A. When isolating DNA from multiple flies, elute in a volume of 500 µl. For single flies, elute in a volume of 300 µl. Other genomic prep kits or protocols may be followed for this step.

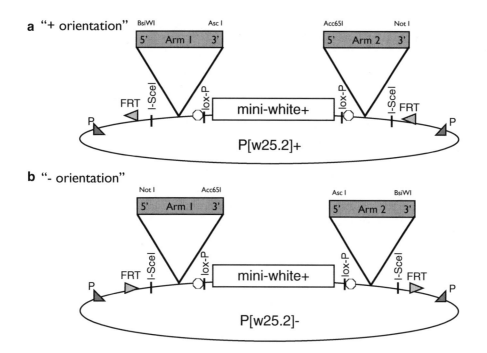

Fig. 5. P[w25.2] vector. (**a**) The P[w25.2] vector was derived from pCasper4 (27). In addition to the P-element inverted repeats, which allow for insertion into the genome, P[w25.2] contains the mini-white eye color marker flanked by lox-P sites and a six-frame stop site sequence. Four unique cloning sites allow directional arm insertion. Two *FRT* sites in the same orientation allow for *hs-FLP* excision of sequence between the two sites creating an extra chromosomal circular DNA molecule. *hs-I-Sce I* then cleaves the circular molecule creating a linear extrachromosomal fragment capable of recombining with the endogenous locus via the two homology arms. The lox-P sites allow for removal of the mini-white sequence via *Cre* recombinase. (**b**) P[w25.2] drawn in reverse (–) orientation. Which orientation to use depends on the restriction sites chosen and the 5′–3′ orientation of the arms being cloned.

2. Amplify both wild-type arms using primers, which include the necessary restriction enzyme cloning sites at the 5′ and 3′ ends (Fig. 1a). Use a high-fidelity Taq polymerase such as Platinum Taq-HF or Phusion Taq to amplify the arms following the manufacturer's instructions. Standard Taq polymerase may be used for this step. However, due to higher error rates, more clones may have to be sequenced to get an error-free product.

3. Check an aliquot of the PCR reaction on a 0.8% agarose gel. If the product is abundant with few or no contaminating nonspecific products, purify DNA from the remaining reaction mixture using the Wizard Gel and PCR Cleanup Kit following the kit instructions. If there are significant contaminating nonspecific PCR products then run the remaining product out on a gel and cut out the gel slice containing the correct band. The gel slice can be processed with the same Wizard Gel and PCR Cleanup Kit following the instructions for gel slice cleanup.

4. The resulting clean PCR product can then be cloned via one of two TOPO Cloning Kits, which allow rapid and efficient cloning of PCR products (Fig. 1b). If the PCR product was generated using a polymerase that leaves a single A nucleotide overhang, use the TOPO-TA kit. If the polymerase creates a blunt-ended product, use the TOPO-Blunt kit. The procedure for both kits is the same. Because Topo cloning is quite efficient we have found that we can reduce the volumes specified in the instructions, thereby allowing more reactions per kit. We follow the manufacturer's instructions except that we scale down the reaction as follows:

Topo cloning

2.4 µl of PCR product + water

0.3 µl of salt solution

0.3 µl of Topo vector

3.0 µl total

5. Following transformation of the cloning reaction, pick ten colonies and inoculate each into 3 ml of LB medium containing 50 µg/ml of ampicillin.

6. Grow overnight in a shaking incubator at 37°C.

7. Purify the plasmid DNA using any standard plasmid miniprep protocol such as the Pure Yield Plasmid Mini-prep Kit (Promega).

8. Check each sample by restriction digest with the appropriate enzymes to confirm the release of correct fragments: ~2.5 Kb insert and ~4 Kb vector.

9. Sequence constructs to identify one clone of each arm that is error free or contains only synonymous changes. These two clones – Topo-arm I and Topo-arm 2 – will be used to create the wild-type P[w25.2] construct, and they will be the templates for subsequent mutagenesis. Figure 1c shows appropriate primers.

3.2. Sequencing

Sequencing protocols vary depending on the resources available at each institution. Many facilities no longer maintain in-house sequencing facilities, as there are many outside service options that are more cost effective. We prefer to do our own reactions and cleanup and then submit the samples for running to an outside vendor (see Note 3). Numerous primers, both gene and vector specific, are used for sequence validation throughout construct prep (Fig. 6a, b) and posttargeting (Fig. 6c, d).

1. Standard Big Dye Sequencing reaction:

Assemble on ice in 0.2 ml thin-walled PCR tubes:

X µl DNA template (~100 ng for plasmids, 30–90 ng for PCR products)

2 µl of primer (10 µM)

a   Arms 1 & 2 in Topo vector

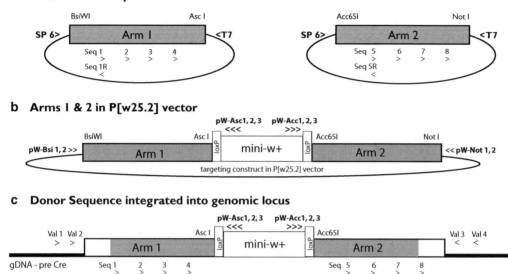

b   Arms 1 & 2 in P[w25.2] vector

c   Donor Sequence integrated into genomic locus

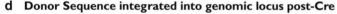

d   Donor Sequence integrated into genomic locus post-Cre

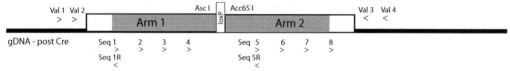

Fig. 6. Diagram of suggested primers for sequencing plasmids and integrated constructs. Vector primers (*bold*) are provided in Subheading 2. Gene-specific primers are designed by user. (**a**) Homology arms 1 and 2 in Topo vector. SP6 and T7 vector primers provided. Five forward sequencing primers (~600 bp apart) and one reverse primer are designed to cover each arm and the 5′ and 3′ junctions. (**b**) Arms in P[w25.2] vector. Vector primers or gene-specific primers may be used for verifying junction sequences. (**c**) Donor sequence integrated into genomic locus. Vector primers Asc 1–3 and gene specific Val 1 and 2 are used to validate arm 1. Acc 1–3 and gene specific Val 3 and 4 are used to validate arm 2. (**d**) Donor sequence integrated into genomic locus post-*Cre*. Post-*Cre* validation can be accomplished using gene-specific Val primers (full-length product ~5 Kb) or in two overlapping fragments (~2.5 Kb) using Val 1 or 2 with Seq 5R and Val 3 or 4 with Seq 4. Engineered arms are represented by shaded boxes and endogenous arms by *white boxes* (**a–d**).

2 μl of Big Dye Buffer

2 μl of Big Dye

X μl of water

20 μl total

Cycling parameters:

1 cycle (hot start):

94°C×2 min.

25 cycles:

96°C×10 s
50°C×5 s
60°C×4 min

2. CleanSeq magnetic bead cleanup

CleanSeq Beads bind the sequencing products to allow removal of unincorporated nucleotides and other contaminants. The beads are stored at 4°C and settle out of solution rapidly. Make sure beads are mixed thoroughly and completely resuspended before pipetting and vortex periodically while working to maintain even suspension.

Following Big Dye cycle sequencing:

1. Add 6 μl CleanSeq beads to a 20 μl Big Dye reaction

2. Add 80 μl 80% ethanol (prepared fresh) and mix *thoroughly* by pipetting (7×)

3. Place on SPRIPlate® 96R and let sit for ~3 min to allow beads to migrate (do not remove tubes from the plate for rest of the procedure)

4. Remove ethanol by pipetting without disturbing the beads

5. Rinse with 150 μl 80% ethanol (pipette into tubes – do not vortex)

6. Remove as much ethanol as possible by pipette (set volume to ~160 μl to remove wash volume plus any volume left from the previous step)

7. Air dry beads (~5 min in hood)

8. Add 40 μl water to elute DNA

9. Submit samples for running (samples are stored at –20°C until ready to ship)

Sequence verified arm 1 and 2 Topo constructs are now ready for site-directed mutagenesis (Subheading 3.4) and for construction of the wild-type P[w25.2] construct (Subheading 3.3).

3.3. P[w25.2] Construct Assembly and Verification

The two wild-type arms are sub-cloned into the P[w25.2] vector sequentially. Cloning order is determined by the arm sequences. If arm 1 carries a restriction site used for cloning arm 2, then arm 2 must be sub-cloned first. If neither arm has internal sites used for cloning the other arm, the cloning order is irrelevant. The same procedure will be followed for all mutant constructs.

1. Digest wild-type Topo clones with the appropriate enzymes to release the arms. Simultaneously, prepare the P[w25.2] vector by digesting with the two enzymes necessary for inserting the first of the two arms.

2. Purify both vector and arm products by agarose gel and extract DNA with the Wizard Gel and PCR Cleanup Kit as before with arm amplification.

3. Quantitate both vector and arms by spectrophotometry or by visualizing an aliquot of each fragment on a gel and comparing

the band intensities to known standards such as the 1-Kb ladder.

4. Set up ligation reactions using T4 DNA ligase following manufacturer's instructions.

5. Transform 1–3 μl of the ligation reaction into competent cells.

6. Plate on LB medium with 50 μg/ml of ampicillin and incubate overnight at 37°C.

7. Pick 6–10 white colonies and inoculate individual 3 ml overnight cultures.

8. Purify plasmid DNA as before and analyze via restriction digest by dropping out the arm with the two enzymes used for cloning.

9. Choose one positive clone to serve as vector for the second round of arm cloning.

10. Digest vector containing first arm with two enzymes necessary for cloning of the second arm (see Note 4).

11. Purify, clean, and quantitate as 2.

12. Set up second ligation with new vector and second arm insert as above.

13. Transform, plate, pick colonies, and purify plasmid DNA as before.

14. Digest to confirm cloning of second arm.

15. Set up three separate digests with a positive clone. BsiWI/Asc I to drop out arm 1, Acc65I/Not I to drop out arm 2, and Sce I to confirm presence of Sce sites. Sce I will excise a ~9.5-Kb fragment containing both homology arms and the mini-white gene.

16. Sequence across all junctions to confirm proper cloning and presence of SceI, lox-P, and FRT sites.

17. After identifying a clone that meets all requirements, re-transform and grow up 50 ml of overnight culture.

18. Purify plasmid DNA with the Pure-Yield Midi Prep kit according to manufacturer's instructions.

19. Determine the DNA concentration and repeat the diagnostic digests (step 14) before submitting DNA for injection. Generally, 50–100 ug of midi-prep quality DNA is sufficient.

3.4. Site-Directed Mutagenesis

Site-directed mutagenesis is performed on one or both Topo-arm constructs to introduce the desired mutation(s). The Quik Change II XL Site-directed Mutagenesis Kit allows efficient generation of point mutations as well as replacement, deletion, or insertion of single or multiple adjacent nucleotides. Briefly, two complementary mutagenic primers are annealed to the denatured template, Topo-arm 1 or 2, and extended via a high-fidelity polymerase creating a mutated plasmid with staggered nicks. Subsequent digest of the reaction with DpnI removes the nonmutated parental template and the remaining reaction is transformed into competent cells where the nicks are repaired. Colonies are prepped as with the initial Topo clones (Subheading 3.1). Digest the plasmids with EcoRI to release the insert to confirm the correct fragment size. Sequence positive clones to identify those carrying the desired mutation. We follow the manufacturer's suggestions with one exception: we do not HPLC or PAGE purify our mutagenic primers unless they are >60 nucleotides in length. Although purified primers do increase the efficiency of mutagenesis, the added cost is probably not necessary. Desired mutations are generally obtained by sequencing ten or fewer clones.

Once sequence-verified mutant Topo clones are identified, mutant constructs are assembled in P[w25.2] as with wild-type clones (Subheading 3.3).

3.5. Generating, Mapping, and Stocking Transgenic Lines

The sequence-verified P[w25.2] constructs are injected into *Drosophila* embryos via standard methods to obtain transgenic lines. This may be done either in-house (23) or through one of several commercially available sources (see Note 5).

Transgenic lines are mapped via the w^+ eye color marker and stocked as homozygotes if possible. Details for mapping and stocking P-element insertions can be found in (25) or the injection service may provide this option. Theoretically, only one transgenic line is needed to achieve multiple successful independent targetings, but it is advantageous to have multiple lines for reasons noted (see Notes 6 and 7).

3.6. Excision of the P[w25.2] Construct to Create the Extrachromosomal Donor Sequence

HR is initiated by crossing a P[w25.2] transgenic line to a line expressing heat shock *FLP* and *Sce* (26) (see Fig. 7 for an overview of the crosses detailed below). It is important that the chromosome being targeted does not carry either the *FLP-Sce* or P[w25.2] transgene (see Note 8). Vials containing eggs and larvae aged 0–48 h undergo heat shock to activate the enzymes *FLP* and *I-Sce I*. *FLP* excises the sequence between the FRT sites. The resulting circular molecule is linearized by *I-Sce I* creating an extrachromosomal donor sequence capable of recombining via the homology arms at the endogenous locus. *FLP* and *I-Sce I* are active in both somatic and germline tissues. The desired targeting event, integration of the targeting construct at the endogenous locus, must occur in the germline in order to be propagated. Germline integration cannot be assayed until the next generation, thus we make use of the somatic activity in surviving adults to

Overview of HR Crossing Scheme

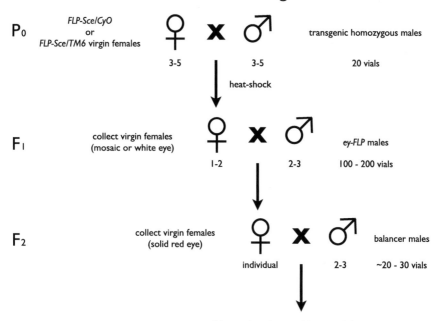

Fig. 7. Overview of HR crossing scheme. P_0: transgenic males are mated to *FLP-Sce* virgin females. Heat-shock induces *FLP-Sce*-mediated excision of the targeting construct creating an extrachromosomal donor molecule capable of recombining at the endogenous locus. F_1: Mosaic or white-eyed females are crossed to *ey-FLP* males to assess potential germline integration events. F_2: Potential integrants (red-eyed females) are mated and then sacrificed for validation PCR assays.

assay the likely level of germline activity. Adult flies are screened for efficient excision of the targeting construct in the soma by following the w^+ eye color marker. Non-excision is scored by solid reddish (w^+) eyes, while excision of the targeting construct results in mosaic or solid white eyes. It is advantageous to make this step as efficient as possible as the desired downstream events are extremely rare (see Note 7).

1. Collect >100 *yw*; *hsFLP, hsSce-1*/CyO or *yw + hsFLP, hsSce-1*/ TM6 virgin females (see Notes 8 and 9).

2. For each transgenic line being targeted, collect 100 homozygous or heterozygous males (see Note 10).

3. Set up 20 vials (numbered 1–20), each containing 3–5 males and 3–5 *hsFLP, hsSce-1*/CyO or *hsFLP, hsSce-1*/TM6 virgin females. All crosses are maintained at room temperature (~22°C) except during heat shock procedure.

4. Let parental flies mate and lay eggs for 3 days. On the third day, pass parental flies to a new set of 20 numbered vials.

5. Take the initial 20 vials containing fertilized eggs, and heat shock in a 38–39°C water bath for 1 h. Repeat heat shock 24 h later (see Note 11).

6. Let parental flies lay eggs for two days in the second set of vials and then pass to new vials on the third day. Heat shock progeny twice as above.

7. Again, let parental flies lay eggs for two days, discard the parental flies, and then heat shock progeny as above. At this point, you will have 60 vials of progeny per transgenic line used, each of which has been heat shocked twice at 38–39°C for 1 h on subsequent days.

8. Screen F_1 progeny for virgin females that carry both *hsFlp*, *hsSce-1*, and the P[w25.2] transgene. If using *hsFLP*, *hsSce-1/ CyO* females and homozygous transgenic males, collect non-Cy (thus carrying the *hsFlp*, *hsSce-1* transgene) virgin females with white/mosaic eye color (indicating excision of the targeting construct from the initial insertion site). If using *hsFLP*, *hsSce-1/TM6* females and homozygous transgenic males, collect non-*Ubx* virgin females with white/mosaic eye color. If using heterozygous transgenic males collect only mosaic females (see Note 11). Collect virgin females until you have 10–20 from each numbered set of vials. At least 100 virgin females per transgene insertion line are required for the next cross.

3.7. Assaying for Potential Integration via ey-Flp Crosses

We can now assay the germline of the F_1 white or mosaic females collected above by mating to *ey-FLP* males. There are three possible outcomes for each egg from the heat-shocked females. If the donor sequence excised but did not integrate then the progeny will have white eyes due to loss of the mini-white gene. If the donor sequence did not excise then the progeny will have solid pigmented (red) eyes due to retention of the mini-white gene. If the donor sequence excised and integrated at the target locus then the progeny will also have solid pigmented eyes due to the presence of mini-white gene in the integrated donor sequence. The latter two classes of pigmented progeny can be distinguished from each other by mating to *ey-FLP*. The *ey-FLP* construct expresses *FLP* recombinase only in the eye. The *FLP* recombinase can excise the mini-white gene (creating white-eyed progeny) only if the flanking FRT sites are present. The FRT sites will still be present if the initial transgene never excised. However, if the transgene excised and integration occurred, the FRT sites will be lost and *ey-FLP* will have no effect. Thus, any red-eyed F_2 female progeny from this cross are candidate flies for the desired integration event.

1. Set up 100–200 single pair crosses of donor females (mosaic or white-eyed collected above) to *ey-FLP* males (see Notes 12 and 13). Number the vials as follows: 1.1–1.3 … for females collected from any of the three parent vials labeled "1."

2. Check vials after 4–5 days and remove any in which mating has been unsuccessful.

3. Discard parental flies from crosses that have been successfully mated when progeny reach the early pupal stage.

4. Screen vials for red-eyed (solid pigment) female progeny. It is preferable to collect females as virgins. However, as this is a rare event, nonvirgin females are not to be discarded.

5. Cross individual red-eyed females (virgin or non-virgin) to 2–3 w^1 males or an appropriate balancer stock (see Note 14). Label vials continuing the numbering from above; if there are two red-eyed females collected from vial 2.3 then the subsequent vials would be labeled 2.3.1 and 2.3.2 (see Note 15).

3.8. Validation of Integration via PCR

Once F_2 females from step 5 have laid a sufficient quantity of eggs and viability is confirmed, genomic DNA can be isolated from the maternal fly. The gDNA can then be used as template for a diagnostic PCR to check for integration using primers designed to the P[w25.2] vector and the flanking genomic sequence outside of arms 1 and 2 (see Fig. 8a and Note 15). If the F_2 female dies after producing larvae but before collection, validation will be performed

Validation PCR

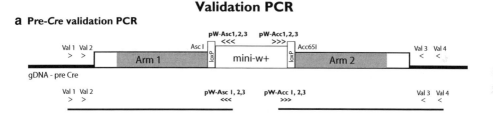

Fig. 8. Validation PCR (**a**) Pre-*Cre* validation PCR. Gene-specific primers flanking arms 1 and 2 paired with vector-specific primers can only generate product (~2.5 Kb) if the integration occurred at the endogenous locus. (**b**) Post-*Cre* validation PCR. After removal of the mini-white gene from the targeted locus, the entire region can be amplified (~5 Kb). The flanking Val primers will amplify from both the endogenous (nontargeted) allele and the targeted allele in heterozygous flies. Product from the targeted allele will cut with both Asc I and Acc65I whereas product from the endogenous (nontargeted) allele will remain intact. This full-length product from the targeted allele should be sequenced to verify the integrity of the integration. Engineered arms are represented by *shaded boxes* and endogenous arms by *white boxes*.

on red-eyed progeny from the next generation. The presence of mutations and/or deletions may be subsequently confirmed through direct sequencing of the genomic locus. If the mutations happen to add or abolish restriction sites, restriction digests may be used as an initial screen prior to sequencing.

3.9. Establishment of Validated Stocks

Validated stocks are maintained for at least three independent lines per construct (see Note 16). The F_2 females used to establish these lines carry the *ey-FLP*-bearing X chromosome and may carry either the *FLP-Sce*-bearing chromosome or the "remnant chromosome." The remnant chromosome is the chromosome that carried the original P[w25.2] transgene. Excision of the donor cassette from the transgene during heat shock leaves behind approximately 1 Kb of vector sequence at the insertion site. All of these chromosomes are undesirable as they could cause significant problems with downstream analyses. Care should be taken to eliminate these chromosomes when stocking targeted alleles. An example of establishing a "clean" stock is shown in Fig. 9.

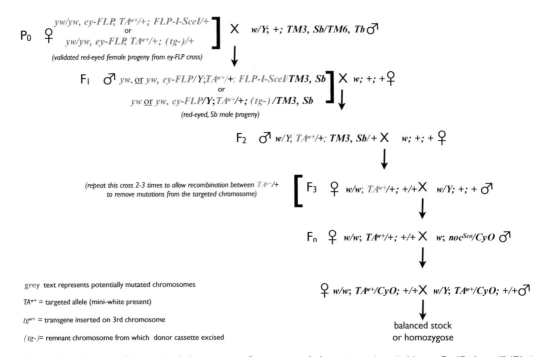

Fig. 9. Removing potentially mutated chromosomes from a second chromosome targeted locus. P_0: "Red-eyed" (TA^{w+}) female progeny from *ey-FLP* crosses are candidates for germline integration events. They may carry either *FLP-I-Scel* or the remnant chromosome (tg^-), which are phenotypically indistinguishable. These females are mated to *w; TM3, Sb/TM6, Tb* males that carry dominant third chromosome markers. F_1: TA^{w+} stubble male progeny are selected and mated to *w* virgin females. The F_2 w^+, nonstubble female progeny will have all potentially mutated chromosomes replaced except for the targeted second chromosome. Repeating this cross several times allows recombination between the TA^{w+} chromosome and its homologue to remove unintended mutations from the targeted chromosome. A balanced stock is created in the F_n by mating females to *w;CyO/noc^{sco}* and selecting red-eyed curly progeny.

3.10. Removing the Mini-White Gene

The mini-white gene within the targeted sequence may potentially cause lethality, alter the expression levels of the gene of interest, or lead to splicing defects. Whether or not any of these conditions hold true for your gene, the ultimate goal is to introduce mutation(s) into the endogenous locus with minimal alteration of that locus. It is therefore necessary to remove the mini-white gene from the targeted lines by crossing to lines expressing *Cre* recombinase. *Cre* recombinase acts at the lox-P sites flanking the mini-white gene and excises the intervening mini-white sequence leaving behind a single lox-P site with flanking six-frame stop codons (Fig. 2 c, d) (26). The lox-P site, flanking stop codons, and the Asc I & Acc65I RE sites used for cloning combine to equal the 76 bp of vector sequence which replaces ~76 bases of non-essential sequence in the target.

1. Set up one or two vials of 3–4 targeted flies mated to 3–4 *Cre* flies. Use targeted males and *Cre* virgin females for targeted loci on the second or third and targeted virgin females with *Cre* males for targeted loci on the X.

2. Transfer parental flies after 3–5 days and discard parentals after an additional 3–5 days.

All F_1 progeny will have white eyes due to the activity of the *Cre* recombinase. Heat shock is not necessary as *Cre* expression is under control of an *Hsp70-Mos 1* promoter, which results in high-level expression in the absence of heat shock (27). Balancer chromosomes with dominant markers are necessary for following the post-*Cre* targeted allele in the absence of the w^+ eye color. There are multiple *Cre* lines with balancers and markers optimized for targeted loci on the X, second, or third chromosomes. For targeted autosomal loci, using females from an X-bearing *yw*, *Cre* line insures that all progeny carry one copy of *yw*, *Cre*. However, not all progeny will receive a copy of the targeted allele if the parental male is heterozygous. If the pre-*Cre* targeted line is lethal due to insertion of the mini-white gene, heterozygous flies must be used for the *Cre* cross. Thus, there are multiple crossing options and outcomes. All options are dealt with separately later and in Figs. 10–12.

3.10.1. Excising the Mini-White Gene from Targeted Second Chromosome Loci

The *Cre* removal of the mini-white gene from a targeted allele on the second chromosome (TA^{w+}) balanced over the dominantly marked *CyO, amos^{Roi-1}* is shown in Fig. 10. There are only three viable F_1 progeny classes (*CyO/CyO, amos^{Roi-}* is lethal). The three classes can be distinguished by straight or curly wings (*CyO*) and smooth or rough eyes (*amos^{Roi-}*). All progeny carry one copy of *yw*, *Cre*. However, only two of the classes, *yw*, *Cre; noc^{Sco}/TA$^-$* and *yw*, *Cre; CyO/TA$^-$* (Fig. 10 F_1 A and B), will also carry the targeted allele, now designated *TA$^-$*. Of the two classes, *yw*, *Cre;*

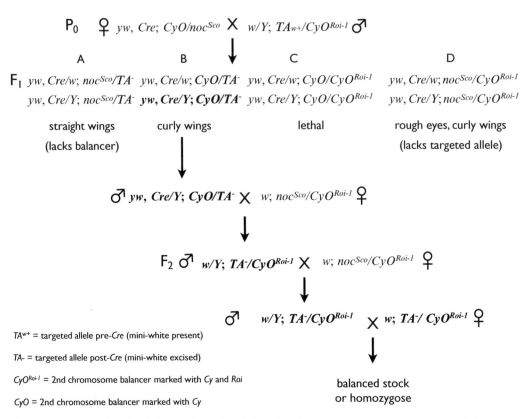

P_0 ♀ *yw, Cre; CyO/noc^{Sco}*   X   *w/Y; TA^{w+}/CyO^{Roi-1}* ♂

| A | B | C | D |

F_1 *yw, Cre/w; noc^{Sco}/TA^-*   *yw, Cre/w; CyO/TA^-*   *yw, Cre/w; CyO/CyO^{Roi-1}*   *yw, Cre/w; noc^{Sco}/CyO^{Roi-1}*

yw, Cre/Y; noc^{Sco}/TA^-   **yw, Cre/Y; CyO/TA^-**   *yw, Cre/Y; CyO/CyO^{Roi-1}*   *yw, Cre/Y; noc^{Sco}/CyO^{Roi-1}*

straight wings   curly wings   lethal   rough eyes, curly wings

(lacks balancer)   (lacks targeted allele)

♂ **yw, Cre/Y; CyO/TA^-**   X   *w; noc^{Sco}/CyO^{Roi-1}* ♀

F_2 ♂ **w/Y; TA^-/CyO^{Roi-1}**   X   *w; noc^{Sco}/CyO^{Roi-1}* ♀

♂   *w/Y; TA^-/CyO^{Roi-1}*   X   *w; TA^-/ CyO^{Roi-1}* ♀

TA^{w+} = targeted allele pre-*Cre* (mini-white present)

TA- = targeted allele post-*Cre* (mini-white excised)

CyO^{Roi-1} = 2nd chromosome balancer marked with *Cy* and *Roi*

CyO = 2nd chromosome balancer marked with *Cy*

balanced stock
or homozygose

Fig. 10. *Cre* removal of mini-white from a targeted allele (lethal) on the second chromosome. P_0: Targeted heterozygous males are mated to *yw, Cre; CyO/noc^{Sco}* females. All potential F_1 classes are shown with identifiable phenotypes listed below each class. Only the relevant genotypic class is shown in the F_2. The genotype to be selected in each generation is in *bold* type. If targeted homozygous males are used in the P_0 cross, then F_1 would have only two progeny classes (*A, B*).

CyO/TA^- is selected as it balances *TA^-* over the dominantly marked *CyO* chromosome. These F_1 males may then be crossed with *w^1; CyO, amos^{Roi-}/noc^{Sco}* females to balance and stock the allele while removing the *yw, Cre* X chromosome.

If the second chromosome targeted allele is viable pre-Cre and the parental male is homozygous (*TA^{w+}/TA^{w+}*) then the crossing scheme would be the same. However, the F1 would only have two progeny classes, A and B (Fig. 10 F_1).

3.10.2. Excising the Mini-White Gene from Targeted Third Chromosome Loci

The *Cre* removal of the mini-white gene from a targeted allele on the third chromosome (*TA^{w+}*) balanced over the dominantly marked *TM6,Tb* is shown in Fig. 11. There are four viable progeny classes in the F_1, which are distinguished from each other by the presence or absence of spread wings (*D**), stubble bristles (*Sb*), and tubby body (*Tb*). Both classes of *TM6, Tb*-bearing flies (Fig. 11 C, D) lack the targeted allele. The remaining two classes, *yw, Cre; D*/TA^-* and *yw, Cre; TM3, Sb/TA^-*, both carry *TA^-* and could be used for subsequent crosses. Of these two, *yw, Cre;*

Removing the mini-white gene from a targeted allele (lethal) on the 3rd

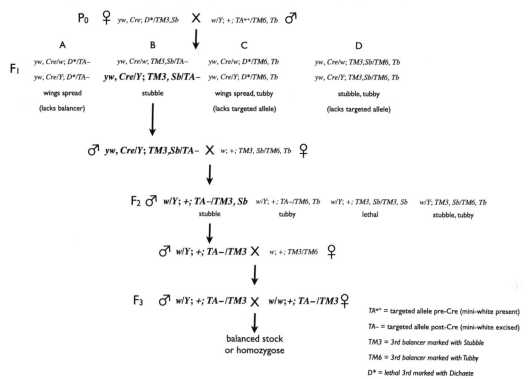

Fig. 11. *Cre* removal of mini-white from a targeted allele (lethal) on the third chromosome. **P₀**: Targeted heterozygous males are mated to *yw, Cre; D'/TM3, Sb* females. All potential **F₁** classes are shown with identifiable phenotypes listed below each class. Only male progeny classes are shown in the **F₂**. There are two potential male progeny classes in the **F₂** that carry the post-*Cre* allele over a balancer. Of the two balancers, *TM3* and *TM6*, *TM3* is generally preferred for ease of scoring the *Sb* marker. The genotype to be selected in each generation is in *bold* type. If targeted homozygous males are used in the **P₀** cross, then **F₁** would have only two progeny classes (*A, B*).

TM3, Sb/TA⁻ males are chosen for the F₂ cross as this balances the post-*Cre* targeted chromosome. The *Cre*-bearing chromosome is removed in the next generation by crossing to *w;TM3, Sb/TM6, Tb* virgin females and selecting F₂ male progeny, which are *TA⁻/TM3, Sb*. By again mating to *w; TM3, Sb/TM6, Tb* virgin females, F₃ male and female progeny are produced which, when mated, yield a balanced post-*Cre* stock.

If the P₀ cross is initiated with a homozygous male, the crossing scheme is the same. Once again, there will be only two F₁ progeny classes, *yw, Cre; D*/TA⁻* and *yw, Cre; TM3, Sb/TA⁻*.

3.10.3. Excising the Mini-White Gene from Targeted X Chromosome Loci

Cre removal of the mini-white gene from a targeted allele on the X chromosome is shown in Fig. 12. If insertion of the mini-white gene in the targeted locus causes pre-*Cre* lethality there will be no *TA^{w+}*-bearing males, thus, targeted females must be used in the P₀ cross (see Note 17). If there is no pre-*Cre* lethality due to the mini-white gene it is still necessary to use targeted females rather

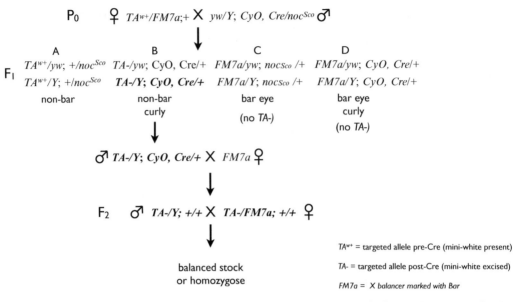

Fig. 12. *Cre* removal of mini-white from a targeted allele (lethal) on the X chromosome. **P₀**: Targeted heterozygous females are mated to *yw; CyO, Cre/noc^Sco* males. All potential **F₁** classes are shown with identifiable phenotypes listed below each class. Only the relevant genotypic class is shown in the **F₂**. The genotype to be selected in each generation is in *bold* type. If targeted homozygous females are used in the **P₀** cross, then **F₁** would have only two progeny classes (*A, B*).

than males for the P_0 cross to produce male F_1 progeny that carry both TA^- and *Cre*. There are four F_1 progeny classes scored by the presence or absence of bar eyes (*FM7a*) and straight or curly wings (*CyO*). Select class B males, $TA^-/Y;CyO, Cre/+$, and mate to *FM7a* virgin females to produce F_2 males and balanced females, which lack the *CyO, Cre* chromosome. Mating F_2 males (TA^-/Y) and females ($TA^-/FM7a$) produces a balanced post-*Cre* stock.

If the P_0 cross is initiated with homozygous targeted females (TA^{w+}/TA^{w+}), only progeny classes A and B will be present in the F_1.

Wild-type and mutant post-*Cre* lines can be maintained as homozygous or balanced stocks. Having the marked balancer present may facilitate downstream crosses for analysis. Some mutations may render post-*Cre* mutant alleles inviable or sterile requiring the targeted chromosome be maintained over a balancer. If wild-type post-*Cre* lines are lethal, it is most likely due to the substitution of the 76 bp of vector sequence for endogenous sequence.

3.11. Backcrossing to Minimize Background Effects

The actions of *FLP*, *Sce*, and *Cre* could potentially introduce unintended mutations in the genomic background that may affect analysis of targeted alleles (28). While the generation of multiple alleles is generally achieved and somewhat circumvents this potential problem, it is recommended, particularly for behavioral analyses, that the recombinant line be backcrossed to a wild-type line for at least five generations in order to remove any unintended mutations that may affect the phenotype of interest. Backcrossing pre-*Cre*, before the mini-white gene is excised, allows the targeted allele to be followed via the w^+ eye color, but it does not address mutations that may be introduced during the *Cre* process. Backcrossing post-*Cre*, in the absence of the w^+ eye color marker requires screening via PCR each generation. Figure 13 shows a sample backcross scheme for a targeted allele on the second chromosome. The P_0 generation begins with a post-Cre targeted allele (TA^-) balanced over *CyO*. F_1 males that are Cy^+ and Sb carry TA^- and will have replaced three of the

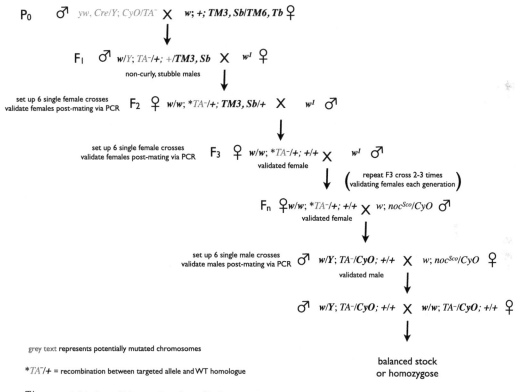

Fig. 13. Sample cross schematic for backcrossing a second chromosome targeted allele to remove unintended mutations caused by the actions of *FLP*, *Sce* and *Cre*. P_0 males carrying a targeted chromosome balanced over *CyO* are mated to *w; TM3, Sb/TM6, Tb* females. In the F_1, $Cy+$, *Sb* males are selected and mated to w^1 females. In the F_2, *Sb* females are selected. All potentially mutated chromosomes, except the second chromosome bearing the targeted allele, have now been replaced. Recombination can occur between the targeted chromosome and its homologue. Females bearing targeted or nontargeted alleles cannot be distinguished, thus individual females are mated to w^1 males. Once larvae are evident, the female is sacrificed for post-*Cre* validation PCR and digests (Fig. 7b). F_3: females (from one validated F_2 cross) are again mated individually to w^1 males and sacrificed for validation once larvae are produced. This process is repeated several times to allow recombination between the targeted chromosome and its homologue to remove any unintended mutations. After multiple generations of backcrossing, a validated female is mated to a male balancer line, *w;nocSco/CyO*. Validated *Cy* male progeny are mated individually back to the balancer line and a balanced or homozygous stock is established.

potentially mutated chromosomes; the X, second, and the marked third. Mating these females to *w¹* males and selecting *Sb+* female progeny results in female progeny where all chromosomes except the one carrying the targeted allele have been replaced and recombination can now occur between the targeted allele and the wild-type homologue. This cross can be repeated as many times as necessary, validating by PCR of each generation, to clean up any unintended mutations on the targeted chromosome. In the final generation, the targeted allele (over the marked balancer *CyO*) is identified by PCR validation and the final post-*Cre* stock is made. Similar crossing schemes can be designed for targeted alleles on the X, third, or fourth chromosome.

This process, while thorough, does not account for mutations that might be tightly linked to the locus being examined. Thus, if only a single wild type or mutant line is examined, even though it has been repeatedly backcrossed, it may still carry mutations that could lead to misinterpretation of phenotypes. By isolating and comparing three independent wild-type targeted alleles and three independent mutant targeted alleles, which have undergone multiple rounds of backcrossing any tightly linked mutations that affect observed phenotypes will be evident and these lines can be eliminated. If multiple independent targeted lines produce similar results in downstream analyses then the observed phenotypic changes can be attributed to the specific mutations genetically engineered via HR.

4. Notes

1. There are many Taq DNA polymerases on the market designed for varying needs. A high-fidelity polymerase is recommended for the initial amplification and cloning steps to minimize PCR artifacts. However, validation PCR does not require a high-fidelity Taq. Many less expensive polymerases generally work quite well for this step.

2. Throughout this chapter we refer to arms 1 and 2 with respect to the published P[w25.2] map that lists the cloning sites in the order BsiWI (arm 1) Asc I and Acc65I (arm 2) Not I. This is drawn as the "+ orientation" (Fig. 5a). It may be necessary, in order to maintain the 5'–3' orientation of the target gene and integrate the correct cloning sites, to turn the map around or reverse the arm designations ("– orientation") as shown in Fig. 5b. Either map is accurate and all vector primers have restriction site labels that apply to either orientation.

3. Most outside services require users to submit template and primer. They perform the reaction and cleanup, run the

sample, and provide users with text file readout. However, sequencing of transcripts to look at RNA editing levels requires that you have access to the actual sequence trace to interpret the mixed A/G peaks. Make sure the vendor you choose will provide actual trace data. The sequencing facility at the University of Wisconsin Biotechnology Center provides options for run only; cleanup and run; or reaction, cleanup, and run (http://www.biotech.wisc.edu/).

4. It is helpful to maintain aliquots of various cloning intermediates to facilitate subsequent cloning steps. For instance, if multiple mutant constructs are planned that only involve mutations in arm 2, it would be useful to maintain a large aliquot of P[w25.2] vector containing arm 1 and predigested for the insertion of the various arm 2 mutants.

5. P[w25.2] constructs transform at a lower efficiency than expected for constructs of this size (~13 Kb). When choosing an outside vendor for injection services, be aware that some vendors will not guarantee transformants with vectors of this size and may charge for additional rounds of injection.

6. P-element insertion is a relatively random event that can occur on any of the four chromosomes, although insertions on the tiny fourth chromosome are rare. We find that insertions are distributed fairly evenly between the X, second, and third chromosomes with a slight bias towards the autosomes. Insertions can cause lethality if the insert disrupts expression of an essential gene. Overall, approximately 30% of insertions show reduced viability or are homozygous sterile/lethal. Generating 6–10 independent transgenic lines per construct generally ensures that you will have 1–3 lines that are homozygous viable and are inserted on a chromosome other than the target chromosome.

7. We have found that some insertions either do not excise efficiently with the FLP recombinase, or the excision event leads to reduced viability of F_1 females leading to lower progeny numbers for subsequent crosses. The higher the percentage of mosaic and white-eyed females, the more efficient the excision, and thus the greater the likelihood of a germline event occurring. If less than 50% of the female progeny bearing both FLP-Sce and the transgene display mosaic or white eyes it is unlikely that germline targeting has occurred. In this case, it is advisable to repeat the initial targeting with a different transgene, or, if there are no other transgenes available, repeat the initial targeting using larger numbers of flies. Adjusting the heat shock regimen is also an option.

8. An undesirable result of the targeting process would be the accidental targeting of the *FLP-I-Sce-1*-bearing chromosome

or the P[w25.2] transgene-bearing chromosome, since the allele of interest would then be genetically linked to a separate P-element insertion, complicating subsequent analysis. Targeting the *hsFLP, hsSce-1* chromosome would also place the targeted allele on a lethal-bearing chromosome preventing analysis of the allele in a homozygous state. For target loci on the X, third, or fourth chromosomes, use the *FLP-I-Sce-1/*CyO stock. For target loci on the second, use the *FLP-I-Sce-1/*TM6 stock. Similarly, it would be undesirable to target the same chromosome from which the donor transgene is excised. Excision of the donor cassette by *FLP* at the FRT sites leaves behind ~1 Kb of vector sequence at the original insertion site, which would again link the targeted allele to a P-element insertion and complicate subsequent analysis. Choose a transgenic line that maps to a chromosome other than the one containing the target locus.

9. We have observed that females from the *Flp-I-Sce-1/*TM6 exhibit partial early adult-stage lethality, with ~20% failing to survive past 1-week posteclosion. Therefore, if using this line, more females will be required than suggested in step A. Females destined for premature lethality can usually be identified by black lesions on the abdomen, which are visible within two days posteclosion. Thus, holding the virgins for 2–3 days prior to mating allows you to eliminate most nonviable females. We have out-crossed the *FLP-I-Sce* third chromosome line into a different genetic background. This has eliminated the partial lethality issue seen in the original line.

10. Homozygous males are preferred as this ensures that all F_1 progeny will carry a transgenic chromosome. If heterozygous males are used, be aware that white-eyed F_1 progeny may be white eyed not because of efficient excision by *FLP* recombinase, but because they never carried a white+-bearing chromosome. In this case, it is best to collect only mosaic virgin females for the subsequent cross.

11. It is important to use a circulating water bath to provide even and consistent temperature to all vials. Placing the vials in the unwire racks maintains space around each vial for water circulation, again aiding in even and consistent heat distribution. Food should be poured to a consistent height of ~3 cm in all vials. Vials must be submerged to a uniform depth such that the water level is ~1 cm above the top level of the food surface. Use large rubber bands to secure the vials in the racks and prevent them from floating.

12. Based on our experience, integration will occur in approximately 3–4% of the vials. However, the frequency of inclusion of your desired mutations will depend, in part, on its location

within the homology arm. Mutations further out in the arm are more likely to be lost during recombination. Therefore, it may be advisable to increase the number of *ey-FLP* crosses in cases where the desired recombination event may be relatively infrequent.

13. We also find there is a high frequency of sterility in the heat-shocked females (20–60%). It is therefore useful, saving screening time and reagents, to set up *ey-FLP* crosses with two virgin females per vial rather than individual females. The likelihood of obtaining two fertile females both with valid integrations events in the same vial is quite remote.

14. Viability of the integrated chromosome depends on both the effect of the mutation and the effect of having the mini-white gene inserted in the target locus. Targeting the wild-type control construct first determines if there is lethality due to insertion of the mini-white gene.

15. Validation PCR is a trial-and-error process. It may be necessary to try multiple primer pairs and even multiple Taq polymerases to find a set that consistently works to amplify both arms 1 and 2. Figure 8a shows two gene-specific validation primers for each arm, Val 1 and 2 to be used with Asc 1–3 for validation of arm 1 and Val 3 and 4 to be used with Acc 1–3 for validation of arm 2. Initial validation using all primer pair combinations, six per arm, will determine which pair works best. Subsequent validations need only use the optimal primer pair for each arm. It is also recommended, once a successful validation is achieved, that it be used as a positive control for subsequent validation PCR.

16. Independent targeting events are defined as follows: targeted events obtained from different transgenes, targeted events from the same transgene obtained from separate experiments, or targeted events from the same transgene that originated in separate vials (the initial set of 1–20 vials from the excision step) from the same experiment. Example: If the suggested numbering regimen is followed and targeted lines are obtained from vials 4.2.1, 4.2.3, and 7.6.1, the first two (4.2.1 and 4.2.3) cannot be considered independent of each other but either is considered independent of the latter (7.6.1).

17. If mini-white gene insertion results in lethality of a targeted locus on the X chromosome there will be no w^+ male progeny from the F_2 cross (Fig. 7). After confirming that this lethality is due only to the presence of the mini-white gene (viability is restored post-*Cre*), this can be used as a prescreening step in future targetings for that locus. Rather than sacrificing potential integrant females after the F_2 mating, wait for progeny to eclose and score for the presence or absence of

red-eyed males. Those lines producing red-eyed males can be eliminated and lines producing only red-eyed females will be used for validation PCR.

Acknowledgments

The authors wish to thank members of the Reenan lab for helpful discussions and suggestions, especially Sarah Goldgar, and Leila Rieder.

References

1. Higuchi, M., Maas, S., Single, F. N., Hartner, J., Rozov, A., Burnashev, N., Feldmeyer, D., Sprengel, R., and Seeburg, P. H. (2000) Point mutation in an AMPA receptor gene rescues lethality in mice deficient in the RNA-editing enzyme ADAR2. *Nature* **406**, 78–81

2. Wang, Q., Khillan, J., Gadue, P., and Nishikura, K. (2000) Requirement of the RNA editing deaminase ADAR1 gene for embryonic erythropoiesis. *Science* **290**, 1765–1768

3. Palladino, M. J., Keegan, L. P., O'Connell, M. A., and Reenan, R. A. (2000) A-to-I pre-mRNA editing in *Drosophila* is primarily involved in adult nervous system function and integrity. *Cell* **102**, 437–449

4. Tonkin, L. A., Saccomanno, L., Morse, D. P., Brodigan, T., Krause, M., and Bass, B. L. (2002) RNA editing by ADARs is important for normal behavior in *Caenorhabditis elegans*. *EMBO J* **21**, 6025–6035

5. Bass, B. L. (2002) RNA editing by adenosine deaminases that act on RNA. *Annu. Rev. Biochem.* **71**, 817–846

6. Gallo, A., Keegan, L. P., Ring, G. M., and O'Connell, M. A. (2003) An ADAR that edits transcripts encoding ion channel subunits functions as a dimer. *EMBO J* **22**, 3421–3430

7. Basillo C, Wahba A. J, Lengyel P, Speyer J. F., and Ochoa S. (1962) Synthetic polynucleotides and the amino acid code. V. *PNAS* **48**, 613–616

8. Hanrahan, C. J., Palladino, M. J., Ganetzky, B., and Reenan, R. A. (2000) RNA editing of the *Drosophila para* Na⁺ channel transcript. Evolutionary conservation and developmental regulation. *Genetics* **155**, 1149–1160

9. Reenan, R. A. (2005) Molecular determinants and guided evolution of species-specific RNA editing. *Nature* **434**, 409–413

10. Hoopengardner, B., Bhalla, T., Staber, C., and Reenan, R. (2003) Nervous system targets of RNA editing identified by comparative genomics. *Science* **301**, 832–836

11. Grauso, M., Reenan, R. A., Culetto, E., and Sattelle, D. B. (2002) Novel putative nicotinic acetylcholine receptor subunit genes, Dalpha5, Dalpha6 and Dalpha7, in *Drosophila melanogaster* identify a new and highly conserved target of adenosine deaminase acting on RNA-mediated A-to-I pre-mRNA editing. *Genetics* **160**, 1519–1533

12. Smith, L. A., Peixoto, A. A., and Hall, J. C. (1998) RNA editing in the *Drosophila* DMCA1A calcium-channel alpha 1 subunit transcript. *J Neurogenet.* **12**, 227–240

13. Semenov, E. P., and Pak, W. L. (1999) Diversification of *Drosophila* chloride channel gene by multiple post-transcriptional mRNA modifications. *J Neurochem* **72**, 66–72

14. Hanrahan, C. J., Palladino, M. J., Bonneau, L. J., and Reenan, R. A. (1999) RNA editing of a *Drosophila* sodium channel gene. *Ann N Y Acad Sci* **868**, 51–66

15. Bhalla, T., Rosenthal, J. J., Holmgren, M., and Reenan, R. (2004) Control of human potassium channel inactivation by editing of a small mRNA hairpin. *Nat Struct Mol Biol* **11**, 950–956

16. Ingleby, L., Maloney, R., Jepson, J., Horn, R., and Reenan, R. (2009) Regulated RNA editing and functional epistasis in *Shaker* potassium channels. *J Gen Phys* **133**, 17–27

17. Rong, Y. S., and Golic, K. G. (2000) Gene targeting by homologous recombination in *Drosophila. Science* **288**, 2013–2018

18. Rong, Y. S., Titen, S. W., Xie, H. B., Golic, M. M., Bastiani, M., Bandyopadhyay, P., Olivera, B. M., Brodsky, M., Rubin, G. M., and Golic, K. G. (2002) Targeted mutagenesis by homologous recombination in *D. melanogaster. Genes Dev* **16**, 1568–1581

19. Wong, S. K., Sato, S., and Lazinski, D. W. (2001) Substrate recognition by ADAR1 and ADAR2. *RNA* 7, 846–858

20. Keegan, L. P., Brindle, J., Gallo, A., Leroy, A., Reenan, R. A., and O'Connell, M. A. (2005) Tuning of RNA editing by ADAR is required in *Drosophila*. *EMBO J* 24, 2183–2193

21. Palladino, M. J., Keegan, L. P., O'Connell, M. A., and Reenan, R. A. (2000) dADAR, a *Drosophila* double-stranded RNA-specific adenosine deaminase is highly developmentally regulated and is itself a target for RNA editing. *RNA* 6, 1004–1018

22. Jones, A. K., Buckingham, S. D., Papadaki, M., Yokota, M., Sattelle, B. M., Matsuda, K., and Sattelle, D. B. (2009) Splice-variant- and stage-specific RNA editing of the *Drosophila* GABA receptor modulates agonist potency. *J Neurosci* 29, 4287–4292

23. Ashburner, M. (1989) *Drosophila*: A Laboratory Handbook. *Cold Spring Harbor Laboratory*, New York

24. Staber C. and Mann R. Quickly and Easily Isolate Genomic DNA from *Drosophila* With No Preprocessing Using the Maxwell(r) 16 Instrument. *Promega Corporation Web site.* http://www.promega.com/pubs/tpub_017.htm Updated February 2010. New York

25. Greenspan, R. J. (1997) Fly Pushing: the theory and practice of *Drosophila* genetics. *Cold Spring Harbor Laboratory*, New York

26. Maggert, K. A., Gong, W. J., and Golic, K. G. (2008) Methods for homologous recombination in *Drosophila*. *Methods Mol Biol* 420, 155–174

27. Siegal, M. L. and Hartl, D. L. (1996) Transgene coplacement and high efficiency site-specific recombination with the Cre/loxP system in *Drosophila*. *Genetics* 144, 715–726

28. O'Keefe, L. V., Smibert, P., Colella, A., Chataway, T. K., Saint, R., and Richards, R. I (2007) Know thy fly. TIGS 23, 238–242

Chapter 4

Laser Microdissection and Pressure Catapulting of Single Human Motor Neurons for RNA Editing Analysis

Hui Sun, Aruna Raja, Mary A. O'Connell, Valerie Mann, Brendon Noble, and Liam P. Keegan

Abstract

Glutamate is the major excitatory neurotransmitter in the mammalian nervous system. The properties of their ionotropic glutamate receptors largely determine how different neurons respond to glutamate. RNA editing in pre-mRNAs encoding subunits of glutamate receptors, particularly the GluR 2 subunit of AMPA receptors, controls calcium permeability, response time, and total ion flow in individual receptors as well as the density of AMPA receptors at synapses through effects on ER assembly, sorting, and plasma membrane insertion. When RNA editing fails in a neuron, calcium influx through AMPA receptors may cause neuron death by glutamate excitotoxicity, as in the case of vulnerable hippocampal CA1 pyramidal neurons that die after transient forebrain ischemia. Elevated cerebrospinal glutamate is common in ALS and loss of *GluR 2* Q/R site RNA editing has been reported to occur selectively in lower motor neurons in a majority of Japanese sporadic ALS patients. We describe our methods for laser microdissection followed by RT-PCR analysis to study RNA editing in single motor neurons.

Key words: RNA editing, ALS, Neurodegeneration, Brain bank, Spinal motorneurons, GluR-B, GluR 2, Laser microdissection and pressure catapulting (LMPC), Cryosectioning, Single cell analysis

1. Introduction

Amyotrophic lateral sclerosis (ALS) is a progressive neurodegenerative disorder characterized by a selective loss of upper and lower motor neurons. Most cases are sporadic and familial cases account for fewer than 10% of all the cases (1). It has been proposed that glutamate excitotoxicity contributes to selective neuronal death in sporadic ALS (2). The pre-mRNA encoding the glutamate receptor subunit B (GluR 2, also referred to as GluR-B), is edited

Ruslan Aphasizhev (ed.), *RNA and DNA Editing: Methods and Protocols*, Methods in Molecular Biology, vol. 718,
DOI 10.1007/978-1-61779-018-8_4, © Springer Science+Business Media, LLC 2011

to 100% at a critical position (Q/R) whereby the genetically encoded glutamine is replaced by an arginine residue as a result of RNA editing by ADAR2 (3). Editing at this position within the ion pore controls the calcium permeability of the receptor as the receptor is impermeable to calcium when arginine is present (4). However, if the *GluR 2* transcript is not edited at this position then the influx of calcium can result in neuronal cell death, as in the case of vulnerable CA1 hippocampal neurons after transient forebrain ischemia in a rat model for stroke (5). Therefore it was hypothesized that editing at this position could cause sporadic ALS or contribute to disease progression. When editing at the Q/R site was investigated in ALS patients, a decrease in editing at the Q/R site was observed (6, 7).

The control of AMPA receptor gating and calcium permeability by the level of expression and the editing status of the *GluR 2* transcript was originally determined by combining electrophysiological measurements on AMPA receptors of a range of neuronal cell types with a second micropipette extraction of nuclear and cytoplasmic contents from the same individual cells for RT-PCR analysis (8). The method requires a high level of specialist equipment and skill. Shin Kwak and coworkers in Tokyo investigated RNA editing in ALS patients by laser dissection of single motor neurons from the spinal cord of postmortem ALS patients and controls (9). The mRNA was extracted and RT-PCR was performed on the sample. A diagnostic restriction digestion was then performed on the products that were analyzed with a Bioanalyser. This method is easier to use than micropipette extraction and is suitable for analysis of frozen human postmortem material from tissue banks.

We describe this method for isolation of single motor neurons from heterogeneous spinal cord sections by noncontact Laser Microdissection and Pressure Catapulting (LMPC) using a PALM MicroBeam system (Carl Zeiss, Germany). This system utilizes a 337-nm pulsed UVa-nitrogen laser (PALM MicroBeam) interfaced with an inverted microscope (Zeiss Axio Observer) and precisely focused through an objective to a beam spot size of less than 1 μm in diameter for the microdissection action. The laser traces a path around the motor neurons that is defined by a software that controls a motorized stage. A polyethylene–naphthalate (PEN) membrane underlying the spinal cord slice absorbs the UV laser light and melts itself and the overlying tissue section along the cutting path (10–12). After microdissection is complete a defined laser pulse catapults the isolated specimen, supported by a section of PEN membrane, out of the object plane directly into a collection device in a noncontact and contamination-free manner. With LMPC the laser pulse is directed at the specimen for ≤1 ns thus minimizing heat transfer.

2. Materials

2.1. Tissue

The method described in this chapter has been used to analyze human postmortem tissue from ALS and control patients obtained from the Sheffield Brain and Spinal Cord Tissue Bank. However, the same method can be used on the spinal cord from either rats or mice. When planning this work the time required to obtain all the relevant ethical approvals has to be taken into consideration as this may consume a significant amount of time. In addition, hepatitis vaccination is advisable when handling human tissue and local Health and Safety Rules for work with human tissue apply (see Note 1).

2.2. Laser Microdissection and Pressure Catapulting of Single Human Motor Neurons Under CL2 Containment

1. PALM® MicroBeam System (P.A.L.M MicroBeam Technologies GmbH, Germany, subsequently acquired by Carl Zeiss)
2. Dry ice
3. Bright 5030 cryostat (chamber temperature –20°C specimen holder –15°C) (Bright, Huntington, UK)
4. Jung tissue freezing medium (Leica Instruments, Germany)
5. RNase-free Membrane Slide 1.0 PEN (polyethylene naphthalate) (Carl Zeiss Microimaging GmbH, Germany)
6. Hair drier
7. Ice-cold 100% methanol
8. 0.1% Toluidine Blue O (Sigma-Aldrich) (dissolve 0.1 g of Toluidine Blue in 100 ml of RNAse-free water and then filter with 0.22 mm filter), store at 4°C
9. Ice-cold 70% ethanol
10. Ice-cold 100% ethanol
11. Silica gel, Type III (Sigma-Aldrich)

2.3. RNA Extraction

1. TRIzol reagent (Invitrogen, Life Technologies)
2. Chloroform
3. Phase-Lock Gel Heavy 1.5 ml (Eppendorf)
4. Phenol chloroform isoamyl alcohol mixture 25:24:1
5. 3 M sodium acetate, pH 5.2
6. Pellet Paint® NF Co-Precipitant (Novagen, Darmstadt, Germany)
7. Ethachinmate (WakoPure Chemical Industries, Ltd., Osaka, Japan)
8. 2-Propanol
9. Ice-cold 75% ethanol
10. RNase-free water (present in Sensiscript RT kit)
11. Heating block at 65°C

2.4. Reverse Transcription and Nested PCR

1. Sterile laminar flow hood
2. Sensiscript RT kit (Qiagen)
3. RNA guard™ RNase Inhibitor (GE Healthcare)
4. Oligo dT (Promega)
5. Heating blocks at 37°C and at 93°C
6. Advantage 2 Polymerase Mix (Clontech)
7. PCR primers
8. hG2F1 (5′-TCTGGTTTTCCTTGGGTGCC-3′)
9. hG2R1 (5′-AGATCCTCAGCACTTTCG-3′)
10. hG2F2 (5′-GGTTTTCCTTGGGTGCCTTTAT-3′)
11. hG2R2 (5′-ATCCTCAGCACTTTCGATGG-3′)

2.5. Restriction Digestion and Analysis

1. 10× restriction buffer
2. Restriction enzyme *Bbv*I (New England Biolabs)
3. Heating blocks at 37°C and at 65°C
4. Agilent 2100 Bioanalyser (Agilent Technologies, Inc., CA, USA)

---

3. Methods

This method uses human tissue; therefore all the legal requirements have to be met prior to the commencement of this project. In addition, local health and safety rules have to be strictly adhered to (see Note 1). As the quality of the mRNA is crucial to the success or failure of this method, effort has to be made to ensure that it does not get degraded at any step. The spinal cord should be frozen on dry ice as soon as it is feasible after removal and stored at –80°C until use.

3.1. Cryostat Sectioning and Toluidine Blue Staining of Spinal Cord Sections Under CL2 Containment for Working with Human Tissues

1. Frozen spinal cord from thoracic segment six (T6) (see Note 2) is placed on dry ice and transferred into a Bright 5030 cryostat (Bright, Huntington, UK) chamber. The chamber temperature should be precooled to –20°C and the specimen holder to –15°C. The spinal cord block is attached to the metal specimen holder block by being overlaid with Jung tissue freezing medium (Leica Instruments, Germany) and frozen quickly on dry ice. Sections at 15 μm thickness are cut from the frozen tissue and mounted onto RNase-free PEN (polyethylene naphthalate) membrane glass slide (Carl Zeiss Microimaging GmbH, Germany) for staining and microdissection. The slides are placed on dry ice and put at 4°C (see Note 3).

2. All subsequent steps until TRIzol extraction are performed at 4°C. The sections are dried with a hair drier on cool setting for approximately 5 min and subsequently fixed for 1 min with 100% methanol. The methanol is then discarded and the section is dried for up to 3 min with a hair drier on the cool setting. The section is then stained with 0.1% toluidine blue for 1 min and the dye is discarded. The section is washed with 70% ethanol and 100% ethanol for 10 s each and again dried with the hair drier on cool setting for 2 min. The slides are then placed in a 50-ml falcon tube containing silica gel and stored at –80°C until required (see Note 4).

3.2. Identification and Laser Microdissection with Pressure Catapulting (LMPC) of Single Motor Neurons Using the PALM MicroBeam System

Single motor neurons are isolated using LMPC PALM MicroBeam System (Carl Zeiss Microimaging GmbH, Germany). This system utilizes a 337-nm pulsed UVa-nitrogen laser (PALM MicroBeam) interfaced with an inverted microscope (Zeiss Axio Observer). LMPC is operator-controlled using PALM RoboSoftware 4.2 (information available at *zeiss.de/microdissection*). Optimal LMPC is obtained using dehydrated tissue specimens and care should be taken to ensure water is eliminated.

1. Turn on the power for microscope and laser control box. There is no warm-up period required for the P.A.L.M.

2. The PEN membrane glass slide containing the prepared and stained tissue specimen is placed on the microscope stage (see Note 5).

3. The LMPC software program is accessed by double click on the PALMRobo software icon.

4. Focus live image on the PC screen, which displays the current image on the microscope.

5. Focus laser. Rotate the objective of the microscope to be used. Fire a laser pulse in the region not including the tissue with small spot button in the PALMRobo PC screen and then adjust the cutting spot to match the chosen small spot.

6. Set the power input and focus the UV laser. Melting of the PEN membrane and the overlying tissue along a cut path is obtained using a wide UV beam that focuses sharply at the membrane and rapidly goes out of focus above this plane so that catapulted tissue above the slide is not damaged. Use the PALMRobo PC software to define a long continuous cutting path in some area of the PEN membrane lacking tissue on the slide. The speed of cutting along this path can be reduced by reducing the rate of movement of the motorized stage to 10% of normal (the repetition rate of the laser does not change). The slower slide movement gives sufficient time to adjust the laser setting during a cut run. While the laser is cutting along this path, adjust the power and focus of the laser to choose an

adequate size of polymer wetting for a cut with sharply defined edges and an absence of visible burn and damage along the edge of the cut. To be systematic, start with a low power level and low focus and gradually increase the power till a wide burn due to a poorly focused high-energy beam is seen; then focus the beam and reduce the power for a clean narrow cut. When the rate of stage movement is later increased to normal the cut may be a bit narrower. The power may need to be increased a little for this reason or if the tissue section to be cut is thick (see Note 6).

7. Locate the motor neurons in the spinal ventral horn for dissection using the live PC screen. A reasonably complete transverse section is needed to locate the correct motor neurons in relation to landmarks of the spinal cord (Fig. 1a). The image can be saved as premicrodissection image (Fig. 1b). Choose an annotation tool to trace a freehand cutting path around the motor neuron at the margin.

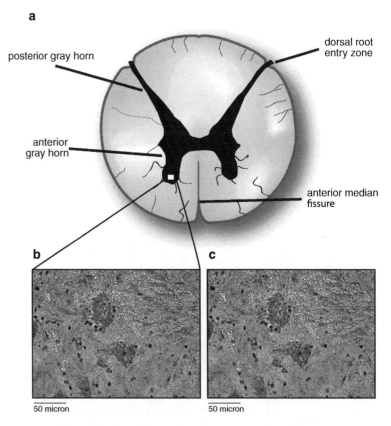

Fig. 1. (a) Cartoon showing the main features of a transverse section of human spinal cord from thoracic segment 6 that guides the identification of motor neurons for dissection. (b) A motor neuron before and (c) after laser-microdissection. The scale bar represents 50 μm.

8. Take a 0.5-ml cap/tube which is purchased as guaranteed as RNase free, pipette 40 μl of TRIzol reagent (Invitrogen, USA) into the middle of the cap, and then put the cap/tube into the cap/tube holder. The image is now seen through the cap and the TRIzol drop so it is no longer very clear.

9. Push the red button on the manual control box to slide the cap/tube holder assembly so that the cap is sitting on the target area for dissection. Then lower down the cap/tube holder assembly onto the slide for efficiently accepting the motor neurons with the cap. Be careful not to touch the tissue slice with the TRIzol reagent or the slice will have to be discarded.

10. Fire the laser to dissect the marginated motor neuron. A closed path around the motorneuron is cut, leaving only a final uncut bridge that holds the motor neuron on the slide. A defined final pulse of larger energy from the UV laser then cuts this bridge and the light force catapults the section into the collection cap. Having the final catapulting laser pulse directed at the periphery rather than in the center of the microdissected piece reduces damage to the cell but sends the fragment spinning rapidly in a direction that does not match the optical axis of the microscope. Therefore the collection cap must be brought close to the tissue slice (see Note 7).

11. After the desired motor neuron has been collected on the cap, lift the cap and push the red button again to rotate the cap/tube holder assembly away from the slide. The image can be saved as a postmicrodissection image (Fig. 1c).

12. Remove the cap/tube from the cap/tube holder, and mix and lyse the cell by inversion, then centrifuge the sample at full speed for a few seconds.

13. After all dissections are completed, shut down the PALMRobo software program and turn off the power to the PALM system.

3.3. RNA Extraction

1. This part of the procedure can be performed in a laminar flow hood to reduce risks of contamination ahead of the RT-PCR amplifications. Two hundred microlitres of TRIzol reagent (Invitrogen, USA) is added to the single neuron in 0.5-ml tube. The samples are stored at room temperature for 5 min and 40 μl of chloroform is added and the tubes are shaken vigorously for 1 min. The tubes are then put on ice for 15 min and then centrifuged at $18,000 \times g$ for 30 min at 4°C.

2. The supernatant is transferred to a Phase-Lock Gel tubes and 120 μl of phenol:chloroform:isoamyl alcohol mixture is added. The Phase-Lock Gel tube is centrifuged at $18,000 \times g$ for 5 min at 4°C before use.

3. The tubes are shaken for more than 10 times followed by centrifugation at $15,000 \times g$ for 15 min at 4°C.

4. The supernatant is removed and placed in a 0.5 ml tube that already contains 10 µl of 3 M sodium acetate and 1 µl of a carrier mixture comprising 0.2 µl of Pellet Paint Co-precipitant NF and 0.8 µl of Ethachinmate. One hundred and ten microlitres of 2-propanol is also added and the mixture is vortexed. The samples are then placed at –20°C for at least 2 h to overnight.

5. The samples are then centrifuged at $18,000 \times g$ for 30 min at 4°C and the supernatant is discarded. The pellet is rinsed with 200 µl of ice-cold 75% ethanol, vortexed for 2–3 s, and then centrifuged at $18,000 \times g$ for 10 min at 4°C and this step is repeated.

6. Subsequently 200 µl of ice-cold 75% ethanol is added to the samples and they are stored at –80°C overnight.

7. Next day the samples are centrifuged at $18,000 \times g$ for 10 min at 4°C and the supernatant is discarded. The pellet is rinsed with 200 µl of ice-cold 75% ethanol, vortexed for 2–3 s, and centrifuged at $18,000 \times g$ for 10 min at 4°C. The samples are air dried for 3–5 min at room temperature.

8. 5.5 µl of RNase-free water is added to the pellet and they are then kept for 2 min at room temperature. The samples are incubated at 65°C for 10 min and subsequently on ice for more than 2 min.

3.4. RT-PCR

RT-PCR should be carried out in a laminar flow hood under sterile conditions to avoid any contamination as two rounds of 35 PCR cycles are performed. The RT reaction is performed with Sensiscript RT kit. In addition, oligo dT and RNA guard ™ RNase inhibitor are required. To the 5.5 µl RNA, add 14.5 µl of the premix containing:

RNase free water	8.17 µl
10× RT Buffer	2 µl
dNTP (5 mM)	2 µl
Oligo dT (10 µM)	1 µl
RNasin (30 U/µl)	0.33 µl
Sensiscript RT	1 µl

The reaction is incubated at 37°C for 2 h and stopped by heating to 93°C for 5 min. The cDNA is stored at –20°C until required.

The first round of PCR is performed with Advantage 2 Polymerase Mix. The final volume is 50 µl and a PCR product is 187 bp.

cDNA from the RT reaction	5 μl
hG2F1 (5′-TCTGGTTTTCCTTGGGTGCC-3′) (10 μM)	1 μl
hG2R1 (5′-AGATCCTCAGCACTTTCG-3′) (10 μM)	1 μl
dNTP mix (10 mM each)	1.25 μl
10× PCR buffer	5 μl
50× Advantage 2 Polymerase mix	1 μl
RNase free water	35.75 μl

PCR conditions

1.	Denaturation step	95°C	1 min	1 cycle
2.	Denaturating	95°C	10 s	
3.	Annealing	64°C	30 s	
4.	Extension	68°C	40 s	
5.	Final extension step	68°C	10 min	1 cycle
6.	Hold	4°C	Forever	

Steps 2–4 are repeated through 35 cycles.

The second round of PCR is a PCR with nested primers that anneal to first PCR product.

PCR is performed with Advantage 2 Polymerase Mix, the final volume is 50 μl and the size of the PCR product is 182 bp.

First PCR product	2 μl
hG2F2 (5′-GGTTTTCCTTGGGTGCCTTTAT-3′) (10 μM)	1 μl
hG2R2 (5′-ATCCTCAGCACTTTCGATGG-3′) (10 μM)	1 μl
dNTP mix (10 mM each)	1.25 μl
10× PCR buffer	5 μl
50× Advantage 2 Polymerase mix	1 μl
RNase free water	38.75 μl

PCR conditions

1.	Denaturation step	95°C	1 min	1 cycle
2.	Denaturating	95°C	10 s	
3.	Annealing	66°C	30 s	
4.	Extension	68°C	40 s	
5.	Final extension step	68°C	10 min	1 cycle
6.	Hold	4°C	Forever	

Steps 2–4 are 35 cycles.

**3.5. Restriction
Digestion Analysis
of PCR Product
for Editing**

To determine the level of editing by ADAR2 at the Q/R site in *GluR 2* a diagnostic digest with *Bbv*I is performed on the PCR product. The editing efficiency is calculated with a 2100 Bioanalyser. If the PCR product has been edited then there are two fragments of 116 and 66 bp (Fig. 2a, c). The unedited transcript fragments are of 35, 81, and 66 bp (Fig. 2a, e). Restriction digestion is performed in a final volume of 20 µl.

Nested PCR product	17 µl
10× Restriction buffer	2 µl
*Bbv*I (2U/µl)	1 µl

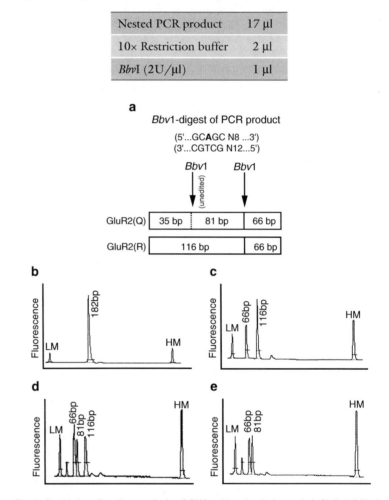

Fig. 2. Restriction digestion analysis of RNA editing levels in pooled *GluR 2* Q/R site RT-PCR products. (**a**) Cartoon showing the locations of the control and the diagnostic *Bbv*I cleavage sites in the RT-PCR product and the fragment sizes generated by cleavage of the unedited *GluR 2*(Q) and the edited *GluR 2*(R) RT-PCR products. The edited A within the asymmetric *Bbv*I site is shown in *bold*. This enzyme cuts many bases away from the recognition sequence. The splice junction between exons 11 and 0 is located 24 base pairs after the edited adenosine. (**b**) Electropherogram generated by a 2100 Bioanalyser of the PCR product of 182 bp prior to *Bbv*I digestion. LM refers to lower marker and HM to higher marker. (**c**) This is an example of 100% editing at the *GluR 2* Q/R site after *Bbv*I digestion showing the presence of only two bands of 66 bp and 116 bp. (**d**) This is an example of 50% editing at the *GluR 2* site after *Bbv*I digestion showing bands of 66, 81, and 116 bp with the 35 bp band not labeled. (**e**) This is an example of no editing at the *GluR 2* Q/R site after *Bbv*I digestion with the presence of the 66 bp and the 81 bp but the diagnostic 116 bp is not present.

The reaction is incubated at 37°C for a minimum of 2 h to overnight and the restriction enzyme is inactivated by heating at 65°C for 30 min.

4. Notes

1. Handling of intact frozen human tissues is in Containment Level 2 (CL2) Safety Cabinets and Cryotomes designated for use of human tissues, wearing lab coat, plastic apron, and disposable rubber gloves. The risk from blood-borne viruses should be reduced in spinal cord. In addition, the volumes of the material used are very low and the use of sharps is avoided. Waste from these hoods and cryotomes including gloves and all materials that have touched the human tissue is retained in the CL2 space for separate disposal. Once cryotome slices of spinal cord have been fixed with methanol they are considered safe to take from the CL2 space to the PALM microscope for dissection of single cells. Cells that have been microdissected into phenol/guanidinium chloride denaturant solution (TRIzol) also do not require CL2 containment as these denaturants also inactivate viruses or prions similar to fixation on microscope slides. This allows the final stages, the RT, and PCR experiments to be carried out in non-CL2 microbiology cabinets and in the normal laboratory.

2. The mid-thoracic T6 segment supplies motor neurons to a very limited set of superficial and deep thoracic muscles and has fewer motor neuron pools than in the thicker lumbar and brachial enlargement regions of spinal cord that also have motor neuron pools supplying a large number of additional arm and leg muscles. Motor neurons are dissected from the anterior horn of the ventral grey matter (large alpha motor neurons of lamina IX supplying striated muscle including cardiac and skeletal muscles). Therefore it is likely that we are looking at neurons from exactly the same motor pool supplying the same muscles in all patient samples.

3. To ensure RNase-free membrane slides, the P.A.L.M Microlaser Technologies suggests dipping the membrane slides into RNase-ZAP (Ambion) for a few seconds, followed by two separate washing in DEPC-treated water and drying at 37°C up to 55°C for 30 min to 2 h. Do not allow the tissue section to dry on the slide at room temperature. Frozen sections should not be stored more for than a few days at –80°C. Freezing should be performed after ethanol fixation or after staining and drying.

4. Complete dehydration of the tissue is necessary for minimizing the upward adhesive forces between the tissue section and the slide.

5. Frozen sections, which are placed in a 50-ml falcon tube containing silica gel, should be dried for approximately 30 min at room temperature before use.

6. It is advisable to store laser settings in a settings file on the hard drive or on a Memory Stick so that the personal settings appropriate to the thickness and the nature of the tissue slices used can be restored if the system has been adjusted by another user.

7. Zeiss sells adhesive collection tubes with dry silicone in the cap. Image formation through the silicone is better than through TRIzol and the dissected fragments captured in silicone remain dry and can be extracted later. Because the cap is dry it can be brought almost into contact with the tissue section, improving the likelihood of capture and avoiding the risk of TRIzol damage to the tissue section.

Acknowledgments

H.S. and A.R. were supported by Grant 6028 from the UK Motor Neurone Disease Association with additional support from the Scottish Motor Neurone Disease Association. L.K and M.O'C are supported by MRC Grant U.1275.01.005.00001.01. We thank Craig Nicol for assistance with figures and Paul Heath, Paul Ince, Pam Shaw and the Sheffield Brain and Spinal Cord Tissue bank for access to the human tissues being used in these studies.

References

1. Cleveland, D. W., and Rothstein, J. D. (2001) From Charcot to Lou Gehrig: deciphering selective motor neuron death in ALS. *Nat Rev Neurosci* 2, 806–819.

2. Shaw, P. J., and Ince, P. G. (1997) Glutamate, excitotoxicity and amyotrophic lateral sclerosis. *J Neurol* 244 Suppl 2, S3–S14.

3. Higuchi, M., Single, F. N., Köhler, M., Sommer, B., Sprengel, R., and Seeburg, P. H. (1993) RNA editing of AMPA receptor subunit GluR-B: A base-paired intron-exon structure determines position and efficiency. *Cell* 75, 1361–1370.

4. Burnashev, N., Monyer, H., Seeburg, P. H., and Sakmann, B. (1992) Divalent ion permeability of AMPA receptor channels is dominated by the edited form of a single subunit. *Neuron* 8, 189–198.

5. Peng, P. L., Zhong, X., Tu, W., Soundarapandian, M. M., Molner, P., Zhu, D., Lau, L., Liu, S., Liu, F., and Lu, Y. (2006) ADAR2-dependent RNA editing of AMPA receptor subunit GluR2 determines vulnerability of neurons in forebrain ischemia. *Neuron* 49, 719–733.

6. Kawahara, Y., Ito, K., Sun, H., Aizawa, H., Kanazawa, I., and Kwak, S. (2004) Glutamate receptors: RNA editing and death of motor neurons. *Nature* 427, 801.

7. Kwak, S., and Kawahara, Y. (2005) Deficient RNA editing of GluR2 and neuronal death in amyotropic lateral sclerosis. *J Mol Med* 83, 110–120.

8. Geiger, J. R. P., Melcher, T., Koh, D.-S., Sakmann, B., Seeburg, P. H., Jonas, P., and Monyer, H. (1995) Relative abundance of

subunit mRNAs determines gating and Ca^{2+} permeability of AMPA receptors in principal neurons and interneurons in rat CNS. *Neuron* **15**, 193–204.

9. Kawahara, Y., Ito, K., Sun, H., Kanazawa, I., and Kwak, S. (2003) Low editing efficiency of GluR2 mRNA is associated with a low relative abundance of ADAR2 mRNA in white matter of normal human brain. *Eur J Neurosci* **18**, 23–33.

10. Vogel, A., Horneffer, V., Lorenz, K., Linz, N., Huttmann, G., and Gebert, A. (2007) Principles of laser microdissection and catapulting of histologic specimens and live cells. *Methods Cell Biol* **82**, 153–205.

11. Vogel, A., Lorenz, K., Horneffer, V., Huttmann, G., von Smolinski, D., and Gebert, A. (2007) Mechanisms of laser-induced dissection and transport of histologic specimens. *Biophys J* **93**, 4481–4500.

12. Schutze, K., and Lahr, G. (1998) Identification of expressed genes by laser-mediated manipulation of single cells. *Nat Biotechnol* **16**, 737–742.

Chapter 5

Biochemical Identification of A-to-I RNA Editing Sites by the Inosine Chemical Erasing (ICE) Method

Masayuki Sakurai and Tsutomu Suzuki

Abstract

Adenosine-to-inosine (A-to-I) RNA editing is a biologically important posttranscriptional processing event involved in the transcriptome diversification. The most conventional method of editing site identification is to compare the cDNA sequence with its corresponding genomic sequence; however, using this method, it is difficult to discriminate between guanosine residue that originated from inosine and errors or noise in the sequencing chromatograms. To address this issue, we developed the inosine chemical erasing (ICE) method to identify inosines in RNA strands utilizing inosine cyanoethylation and reverse transcription PCR. Since this method requires only a limited quantity of total RNA, it can be used in the genome-wide profiling of A-to-I editing sites in tissues and cells from various organisms, including clinical specimens.

Key words: A-to-I RNA editing, ADAR, Inosine, Cyanoethylation, Acrylonitrile, Reverse transcription, PCR, Direct sequencing, Inosine chemical erasing (ICE)

1. Introduction

The posttranscriptional conversion of adenosine to inosine (A-to-I) editing is an RNA modification found abundantly in metazoans from worms to humans (1). Adenosine deaminases that act on RNA (ADARs) catalyze the hydrolytic deamination of adenosines to inosines in double-stranded RNA regions. ADAR activity is required for normal vertebrate development (2–4) and normal invertebrate behavior (5). In addition, A-to-I RNA editing has various functions (6) involved in several neurological disorders (7).

Ruslan Aphasizhev (ed.), *RNA and DNA Editing: Methods and Protocols*, Methods in Molecular Biology, vol. 718, DOI 10.1007/978-1-61779-018-8_5, © Springer Science+Business Media, LLC 2011

1.1. Conventional Method to Identify Inosines on RNA Strand

The most conventional A-to-I editing site identifying method is to compare the cDNA sequence with its corresponding genomic sequence by direct sequencing (8, 9). As inosine can base-pair with cytidine (C), inosines (I) are replaced by guanosines (G) in the cDNA (I-to-G replacement) by reverse transcription (RT) and PCR. Therefore, if adenosine (A) in the genomic sequence is partially or completely replaced with G at the corresponding site in the cDNA sequence, then those sites are candidates for A-to-I editing. However, this traditional method of editing site identification also has several limitations. To exclude possible differences due to single nucleotide polymorphisms (SNPs), the total RNA and genomic DNA must be obtained from the same tissues or cells. It is difficult to discriminate I-to-G replacements in the sequence chromatograms from G contaminations derived from the amplification of pseudogenes or PCR errors. Therefore, the traditional method only detects the I-to-G replacements without excluding the possibility of other G contaminations, thus never unambiguously proves their existence.

Alternatively, A-to-I editing sites can be determined from the decrease of the G peak ratios in the cDNA chromatograms after ADAR knockdown or knockout (10). However, ADAR down-regulation and knockout not only cause changes in gene expression profile and phenotype, but sometimes also result in embryonic lethality (2–4). Furthermore, this method does not allow the detection of inosines in a small amount of clinical specimens. Therefore, identifying the naturally occurring A-to-I editing sites using conventional methods has been difficult.

1.2. Inosine Chemical Erasing Method

To further understand the functional roles of A-to-I editing, we developed a practical and reliable method for direct inosine site identification. Here we describe a new biochemical method, inosine chemical erasing (ICE), to directly and accurately identify inosines in RNA strand, utilizing inosine cyanoethylation and RT-PCR (11). The ICE method enables the discrimination of G residues originating from I-to-G replacements from those originating from allelic SNPs, pseudogenes, and/or sequencing errors. It requires neither the use of genomic DNA, nor the total RNA from ADAR-inactivated cells. The ICE method requires only a limited quantity of total RNA (50–100 ng per A-to-I edited region) from any tissue or cell, including clinical specimens. In addition, the A-to-I editing sites in any RNA molecule, including miRNA, intronic RNA, or mRNA-like noncoding RNA, can be characterized if a primer set can be designed for the target region.

Here we describe our ICE protocol and demonstrate its efficacy using known editing sites in murine mRNA.

2. Materials

2.1. Total RNA Preparation

1. miRNeasy Mini kit (QIAGEN).
2. RNeasy MinElute Cleanup Kit (QIAGEN).
3. RNase-free water.
4. 3 M NaOAc, pH 5.0.
5. Ethanol.
6. Glycogen (2 mg/ml; Roche).

2.2. RNA Cyanoethylation

1. Cyanoethylation buffer: 50% ethanol, 1.1 M triethylammo-niumacetate, pH 8.6. Store at 4°C in a tightly sealed tube.
2. 15.2 M Acrylonitrile. Store at 4°C in the dark.

2.3. RT-PCR

1. 2× Reaction buffer (Invitrogen).
2. SuperScript III RT/Platinum Taq Mix (Invitrogen).
3. 10× PCR buffer for KOD-Plus (Toyobo).
4. dNTPs (2 mM).
5. $MgSO_4$ (25 mM).
6. KOD-Plus (1.0 U/μl; Toyobo).
7. Nusieve 3:1 agarose (Takara).

2.4. Direct Sequencing

1. ExoSAP-IT (GE Healthcare).
2. BigDye sequencing buffer (Applied Biosystems).
3. BigDye v3.1 Ready reaction mix (Applied Biosystems).
4. 125 mM EDTA.
5. Genetic Analyzer 3730XL (Applied Biosystems).

2.5. Genomic PCR

1. 10× PCR buffer for Platinum Taq DNA polymerase (Invitrogen).
2. dNTPs (10 mM).
3. $MgCl_2$ (50 mM).
4. Platinum Taq DNA polymerase (5 U/μl; Invitrogen).

3. Methods

Inosine cyanoethylation in a yeast tRNA anticodon reportedly forms N^1-cyanoethylinosine (ce^1I) through a type of Michael addition with acrylonitrile (12) (Fig. 1a). The N^1-cyanoethyl

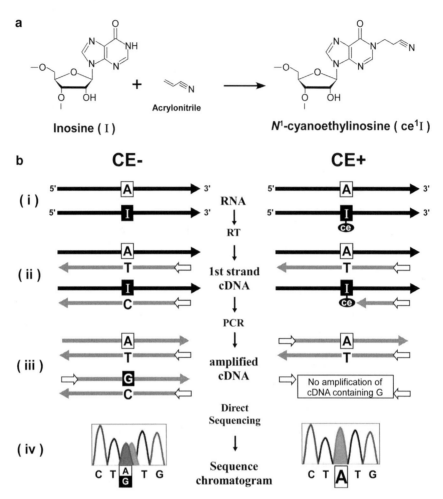

Fig. 1. Inosine cyanoethylation and inosine chemical erasing (ICE) method schematic. (**a**) The inosine (I) cyanoethylation reaction. Acrylonitrile is adducted to the N^1 position of I to form N^1-cyanoethylinosine (ce^{1}I). (**b**) Outline of the ICE method. Schemes without (CE–) or with (CE+) RNA cyanoethylation are shown on the *left* and *right*, respectively. RNA and cDNA are indicated by *black* and *gray arrows*, respectively. (**i**) RNA cyanoethylation. The I residue in the RNA strand is specifically cyanoethylated to form ce^{1}I (CE+). (**ii**) First-strand cDNA synthesis. RNAs are reverse-transcribed by a reverse primer (*opened left arrow*). RNA bearing an A at the editing site is converted to a T in the cDNA in both conditions (CE– or CE+). In the CE– condition, an I residue in the RNA is transcribed to C in the cDNA. In the CE+ condition, first-strand cDNA extension is arrested at the ce^{1}I site. (**iii**) PCR. The first-strand cDNAs bearing a T or C at the editing site are PCR-amplified using a primer set (*open arrows*). The cDNAs bearing C at the editing site are not amplified under the CE+ condition, resulting in the erasure of cDNAs containing G originating from A-to-I editing from the PCR products. (**iv**) Direct sequencing step. In the CE– condition, the editing site is detected as mixed signals of A and G in the sequence chromatogram. In the CE+ condition, the signal of G resulting from A-to-I editing disappears and is replaced mainly with A.

group of ce^{1}I inhibits Watson–Crick base-pairing with C. This principle forms the basis of our ICE method.

The ICE method is composed of four steps (Fig. 1b): (1) RNA cyanoethylation, (2) cDNA synthesis by RT, (3) PCR amplification, and (4) direct sequencing. Total RNA is first either treated with acrylonitrile to cyanoethylate inosines in RNA strands, or left nontreated as a control. Total RNAs are then subjected to first-strand cDNA synthesis by RT. If an A residue is partially edited to an I residue, then the nonedited A and edited I are converted to thymidine (T) and C, respectively, when nontreated total RNA is used as a template for cDNA synthesis. In the cyanoethylated RNA, ce¹I blocks cDNA synthesis, and the cDNA bearing a C residue converted from the edited I does not extend. The cDNAs are then PCR-amplified and sequenced. The A-to-I editing sites in nontreated RNAs are detected as mixed signals of A and G in the sequence chromatogram. However, in cyanoethylated RNAs, the G residues that originated from I residues disappear, and the editing sites are no longer mixed but are replaced mainly with A residues.

3.1. Total RNA Preparation

1. Isolate total RNA from tissues or cells using miRNeasy kit (see Note 1).

2. Precipitate the RNA with 0.1 volume of 3 M NaOAc, pH 5.0, and 2.8 volume of ethanol. If the RNA concentration is assumed to be <40 ng/µl, add 0.02 volume of glycogen (2 mg/ml). Incubate the sample for 30 min at –80°C and centrifuge at $20,000 \times g$ for 15 min at 4°C. Rinse the pellet with 1 ml of 80% ethanol and dry.

3.2. RNA Cyanoethylation

1. Dissolve the RNA pellets with RNase-free water and adjust the concentration to 0.5–10 µg/µl. Transfer 2.5–10 µg of the RNA into a 1.5-ml tube, and adjust the RNA volume to 4 µl with water.

2. Add 30 µl of the cyanoethylation buffer and 4 µl of 15.2 M acrylonitrile to the above solution and mix thoroughly (final concentration of acrylonitrile is 1.6 M). Include a control reaction (CE–) in which 4 µl of the cyanoethylation buffer is used in place of the acrylonitrile. Keep the tube on ice until the next incubation step.

3. Incubate the tubes at 70°C for 15 min for mild cyanoethylation (CE+) and for the control reaction (CE–), or for 30 min for strong cyanoethylation (CE++) (see Note 2).

4. Put the tube immediately on ice and quench cyanoethylation by adding 162 µl of water.

5. Purify the cyanoethylated RNA using the RNeasy MinElute Cleanup kit (see Note 3).

6. Precipitate the RNA as described in Subheading 3.1, step 2.

3.3. Primer Design

1. Two primers sets (for the first and second PCR) should be designed to amplify a region of 300–500 bp, including target inosine sites (see Note 4). The primer set for the nested second PCR should be designed inside the region amplified by

the first PCR. Adjust the Tm value of each primer to ~62°C by the nearest-neighbor method. Choose the primer 3′-ends carefully such that primer–dimer formation is avoided. Each of the second PCR primers is also used for sequencing.

3.3.1. First RT-PCR

1. Dissolve the cyanoethylated total RNA (CE+ and CE++) or nontreated total RNA (CE–) in water and adjust the concentration to 10 ng/μl (CE–), 25 ng/μl (CE+), or 50 ng/μl (CE++) (see Note 5).

2. Transfer 1 μl of the RNA into a PCR tube, and add 1.3 μl each of the first forward primer and first reverse primer (each 2 μM), 2.1 μl of water, 6.3 μl of 2× Reaction buffer, and 0.5 μl of SuperScript III RT/Platinum Taq Mix. Gently mix and flush the solution.

3. Run the RT-PCR reaction as follows: 55°C for 30 min, 94°C for 2 min, totally 10 touch-down cycles from 68 to 62°C (94°C, 30 s; X°C, 30 s; 72°C, 30 s/X=68°, 2 cycles; 66°, 2 cycles; 64°, 3 cycles; or 62°, 3 cycles), followed by 20 normal cycles (94°C, 30 s; 60°C, 30 s; 72°C, 30 s). Keep the samples at 4°C.

3.3.2. Second PCR

1. Dilute 2 μl of the first PCR product mixture with 158 μl water.

2. Transfer 1 μl of the diluted sample into a PCR tube and add 1.9 μl each of the second forward primer as well as second reverse primer (each 2 μM), 4.65 μl of water, 10× PCR buffer for KOD-Plus, 1.25 μl of 2 mM dNTPs, 0.3 μl of 25 mM $MgSO_4$, and 1 μl of KOD-Plus (1 U/μl). Gently mix and flush the solution.

3. Run the PCR reaction as follows: 94°C for 2 min, followed by 25 cycles (94°C, 15 s; 60°C, 30 s; and 68°C, 30 s). Keep the samples at 4°C.

4. Electrophorese 4 μl of the amplified products on a 2% Nusieve 3:1 agarose gel to determine whether the cDNA product migrates as a single band.

3.3.3. Direct Sequencing

1. Mix 5 μl of the second PCR product with 2 μl of ExoSAP-IT, and incubate for 30 min at 37°C. Stop the reaction by heating at 80°C for 15 min.

2. Dilute the sample with 14 μl of water.

3. Transfer 4 μl of the sample into a PCR tube, and add 3.2 μl of sequencing primer (either of the second forwarded primer or reverse primer; 2 μM), 1.9 μl water, 1 μl of 5× BigDye sequencing buffer, and 0.3 μl of BigDye v3.1 Ready reaction mix. Gently mix and flush the solution (see Note 6).

4. Run the sequencing reaction as follows: 96°C for 30 s followed by 25 cycles of PCR (96°C, 10 s; 50°C, 5 s; and 60°C, 4 min).

5. Add 5 μl of 125 mM EDTA to the reaction product, mix thoroughly, and incubate for 10 min at room temperature. Add 25 μl of ethanol, mix thoroughly, and incubate for 10 min at room temperature. Centrifuge at $1,650 \times g$ for 15 min at 4°C. Remove the supernatant, add 5 μl of water and 5 μl of 125 mM EDTA, and mix thoroughly. Add 25 μl of ethanol, mix thoroughly, and incubate for 15 min at room temperature. Centrifuge at $1,650 \times g$ for 15 min at 4°C. Remove the supernatant and add 50 μl of 70% ethanol. Centrifuge at $1,650 \times g$ for 5 min at 4°C. Dry the sample for 10 min at 80°C.

6. Dissolve the sample with 10 μl formamide and boil for 5 min at 95°C. Incubate on ice for at least 2 min.

7. Analyze the samples by a Genetic Analyzer 3730XL. The injection voltage and injection time are set at 1.1 kV and 10 s, respectively, so that the peak signal in the chromatogram is not saturated (see Note 7).

3.3.4. Detection of Inosines in the Sequence Chromatogram

1. Align the obtained sequence with the corresponding genomic sequence from a public database.

2. Search the chromatograms for sites where A is partially or completely replaced with G, but whose corresponding genomic sequence is A (see Note 8).

3. Inosine sites are identified by decreased or erased G peaks relative to A peaks in response to the cyanoethylation strength.

4. In the CE+ and CE++ conditions, if no cDNA amplification is observed, then those sites are considered to be complete editing sites (100% editing ratio) or multiple editing sites. Detailed location of the complete editing sites can be confirmed by the conventional method (Subheading 3.4). Multiple editing sites can be tested by comparing with a weaker cyanoethylation condition (see Note 9).

3.4. Genomic PCR

In cases with complete editing sites (~100%), even after cyanoethylation, the G peak ratio will not decrease. Instead, the cDNA amount after PCR amplification is clearly decreased (Fig. 2a, Mouse Gria2 Q-to-R editing sites), and comparison with the genomic sequence enables determination of the detailed editing site locations.

1. Adjust the concentration of the genomic DNA, obtained from the same tissues or cells as the sample total RNA, to 25 ng/μl.

2. Transfer 1 μl of the genomic DNA into a PCR tube and add 1.3 μl each of the first forward primer as well as first reverse primer (2 μM each), 7 μl water, 1.25 μl of 10× PCR buffer

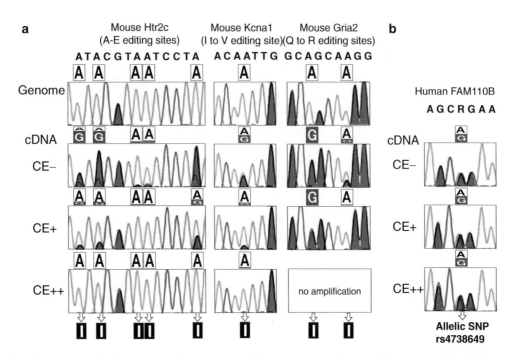

Fig. 2. Inosines erased in the sequence chromatograms. (a) Sequence chromatograms of three known regions bearing A-to-I editing sites: mouse serotonin receptor 2c (Htr2c) mRNA (9), mouse potassium channel Kv1.1 (Kcna1) mRNA (13), and the mouse glutamate receptor B (Gria2) mRNA Q-to-R site (14). *Upper panels* show chromatograms of these regions amplified from mouse genomic DNA. The *second, third,* and *bottom panels* show the chromatograms of cDNAs amplified from mouse brain total RNA treated without cyanoethylation (CE−) or under mild (CE+) or strong (CE++) cyanoethylating conditions, respectively. The G peaks resulting from A-to-I editing were decreased or eliminated proportional to the cyanoethylation strength. Because the Q-to-R site is completely edited and most of the inosine is cyanoethylated in the CE++ condition, no amplification was observed at the Q-to-R site of Gria2 mRNA. (b) Sequence chromatograms of the region bearing the allelic single nucleotide polymorphism (SNP) (rs4738649) in FAM110B mRNA. *Upper, middle,* and *bottom panels* show chromatograms of cDNAs amplified from human total RNA treated without cyanoethylation (CE−) or under mild (CE+) or strong (CE++) cyanoethylating conditions, respectively. No change in A or G ratio at the SNP site was observed, even in CE++ conditions. The sizes of the boxed "A" and "G" reflect their ratios, as estimated from the peak intensity in the chromatograms. G peaks are filled in *gray.*

for Platinum Taq DNA polymerase, 0.25 μl of 10 mM dNTPs, 0.31 μl of 50 mM MgCl$_2$, and 0.05 μl of Platinum Taq DNA polymerase (5 U/μl). Gently mix and flush the solution.

3. Run the PCR reaction as follows: 94°C for 2 min, totally 10 touch-down cycles from 68 to 62°C (94°C, 30 s; X°C, 30 s; 72°C, 30 s/X=68°, 2 cycles; 66°, 2 cycles; 64°, 3 cycles; or 62°, 3 cycles), followed by 20 normal cycles (94°C, 30 s; 60°C, 30 s; 72°C, 30 s). Keep the samples at 4°C.

4. The second PCR (Subheading 3.3.3) is optional. Direct sequencing of the product is described in Subheading 3.3.4.

3.5. Demonstration of the ICE Method

We chose known editing sites in mouse mRNAs for Htr2c (serotonin receptor 2c), Kcna1 (potassium channel Kv1.1), and Gria2 (glutamate receptor) to demonstrate the ICE method (Fig. 2a).

In the nontreated condition (CE–), the editing sites in Htr2c and Kcna1 mRNAs were observed as mixed signals of A and G residues in the cDNA sequence chromatograms, as these sites are partially edited (9, 13). In contrast, the Gria2 mRNA Q-to-R site was observed as a single G peak in the cDNA, because this site is completely edited (14). As a control for each editing site, we also sequenced PCR products amplified from the genomic DNA and confirmed the presence of an A residue at each site.

The G peaks resulting from A-to-I editing were decreased or eliminated proportional to the cyanoethylation strength. In the mild condition (CE+), the ratio of G peaks resulting from I residues decreased in the editing sites of both the Htr2c and Kcna1 mRNAs. In contrast, in the strong cyanoethylation condition (CE++), the G peaks in the editing sites had almost disappeared and were replaced with A peaks. In Gria2 mRNA in the mild condition (CE+), no decrease in the G peak ratio was observed at the ~100% edited Q-to-R site, while the G peak at a partial editing site located 4 nt downstream of the Q-to-R site had disappeared. However, in the strong condition (CE++), no cDNA amplification was detected, because of a complete editing at the Q-to-R site.

Finally, our ICE method could discriminate in cDNAs between I-to-G replacements and other G substitutions, such as allelic SNPs, unfavorable pseudogene amplifications, and/or sequencing errors. For example, we observed no change in the A/G ratio at an allelic SNP (rs4738649), even in the strong cyanoethylation condition (CE++) (Fig. 2b).

4. Notes

1. Purchased total RNA should be repurified with the RNeasy MinElute Cleanup kit. Skip this step if the target RNA is shorter than 200 nt.

2. Seal the tube with Parafilm. After 15-, 30-, and 60-min incubation, ~40, 60, and 80%, respectively, inosines are converted to ce^{1}I.

3. Skip this step if the target RNA is shorter than 200 nt.

4. A-to-I RNA editing sites are frequently found in double-stranded regions formed by repetitive elements. To design specific primers for the target regions, we recommend that the primers to be designed outside the repetitive elements. The secondary structure of the target region, which can be predicted by programs such as CentroidFold (15, 16), RNAz (17), and mfold (18), will help in primer design.

5. The cDNA amplification efficiency usually decreases to some extent proportional to the cyanoethylation strength, probably

because of the nonspecific cyanoethylation of uridines (12). As this decrease is uniformly and independent of the RNA species, it has no effect on the identification of inosines. To compensate for these differences in amplification efficiency, we adjusted the amount of cyanoethylated RNA at each condition.

6. It is sometimes difficult to obtain a clear sequence chromatogram. It is best to sequence the region from both directions.

7. Detecting clear chromatograms is a key success factor in this method. However, our recommended sequencing reagents and instruments can be replaced using other protocols.

8. The G peak ratio in the sequence chromatogram obtained here correlated well with the ratio of cDNA templates bearing G at the editing site. The relationship was linear, and the coefficient of determination (R^2) was >0.96 for cDNA templates larger than 0.001 atto mol. Quantification of the G peak ratio from sequence chromatograms is a reliable way to quantify the frequency of A-to-I editing, assuming that the experiment is performed under the conditions described.

9. In the weaker condition, the total RNA is cyanoethylated with 0.4 M acrylonitrile for 15 min at 70°C.

Acknowledgments

We are grateful to the Suzuki lab members, including Takanori Yano, Hiroki Ueda, Hitomi Kawabata, and Shunpei Okada, for their computational and experimental assistance and fruitful discussions on this study. This work was supported by Grants-in-Aid for Scientific Research on Priority Areas from the Ministry of Education, Science, Sports, and Culture of Japan, and by a grant from the New Energy and Industrial Technology Development Organization (NEDO) (to T. S.).

References

1. Bass, B. L. (2002) RNA editing by adenosine deaminases that act on RNA. *Annu Rev Biochem 71*, 817–846.

2. Higuchi, M., Maas, S., Single, F. N., Hartner, J., Rozov, A., Burnashev, N., Feldmeyer, D., Sprengel, R., and Seeburg, P. H. (2000) Point mutation in an AMPA receptor gene rescues lethality in mice deficient in the RNA-editing enzyme ADAR2. *Nature 406*, 78–81.

3. Wang, Q., Khillan, J., Gadue, P., and Nishikura, K. (2000) Requirement of the RNA editing deaminase ADAR1 gene for embryonic erythropoiesis. *Science 290*, 1765–1768.

4. Wang, Q., Miyakoda, M., Yang, W., Khillan, J., Stachura, D. L., Weiss, M. J., and Nishikura, K. (2004) Stress-induced apoptosis associated with null mutation of ADAR1 RNA editing deaminase gene. *J Biol Chem 279*, 4952–4961.

5. Jepson, J. E., and Reenan, R. A. (2008) RNA editing in regulating gene expression in the brain. *Biochim Biophys Acta 1779*, 459–470.

6. Nishikura, K. (2006) Editor meets silencer: crosstalk between RNA editing and RNA interference. *Nat Rev Mol Cell Biol 7*, 919–931.

7. Maas, S., Kawahara, Y., Tamburro, K. M., and Nishikura, K. (2006) A-to-I RNA editing and human disease. *RNA Biol 3*, 1–9.

8. Paz, N., Levanon, E. Y., Amariglio, N., Heimberger, A. B., Ram, Z., Constantini, S., Barbash, Z. S., Adamsky, K., Safran, M., Hirschberg, A., Krupsky, M., Ben-Dov, I., Cazacu, S., Mikkelsen, T., Brodie, C., Eisenberg, E., and Rechavi, G. (2007) Altered adenosine-to-inosine RNA editing in human cancer. *Genome Res 17*, 1586–1595.

9. Burns, C. M., Chu, H., Rueter, S. M., Hutchinson, L. K., Canton, H., Sanders-Bush, E., and Emeson, R. B. (1997) Regulation of serotonin-2C receptor G-protein coupling by RNA editing. *Nature 387*, 303–308.

10. Nishimoto, Y., Yamashita, T., Hideyama, T., Tsuji, S., Suzuki, N., and Kwak, S. (2008) Determination of editors at the novel A-to-I editing positions. *Neurosci Res 61*, 201–206.

11. Sakurai, M., Yano, T., Kawabata, H., Ueda, H., and Suzuki, T. (2010) Inosine cyanoethylation identifies A-to-I RNA editing sites in the human transcriptome. *Nat Chem Biol 6*, 733–740.

12. Yoshida, M. and Ukita, T. (1968) Modification of nucleosides and nucleotides. VII. Selective cyanoethylation of inosine and pseudouridine in yeast transfer ribonucleic acid. *Biochim Biophys Acta 157*, 455–465.

13. Hoopengardner, B., Bhalla, T., Staber, C., and Reenan, R. (2003) Nervous system targets of RNA editing identified by comparative genomics. *Science 301*, 832–836.

14. Sommer, B., Kohler, M., Sprengel, R., and Seeburg, P. H. (1991) RNA editing in brain controls a determinant of ion flow in glutamate-gated channels. *Cell 67*, 11–19.

15. Sato, K., Hamada, M., Asai, K., and Mituyama, T. (2009) CENTROIDFOLD: a web server for RNA secondary structure prediction. *Nucleic Acids Res 37*, W277–W280.

16. Hamada, M., Kiryu, H., Sato, K., Mituyama, T., and Asai, K. (2009) Prediction of RNA secondary structure using generalized centroid estimators. *Bioinformatics 25*, 465–473.

17. Washietl, S., Hofacker, I. L., and Stadler, P. F. (2005) Fast and reliable prediction of noncoding RNAs. *Proc Natl Acad Sci U S A 102*, 2454–2459.

18. Zuker, M. (2003) Mfold web server for nucleic acid folding and hybridization prediction. *Nucleic Acids Res 31*, 3406–3415.

Part III

Cytidine to Uracil RNA and DNA Editing

Chapter 6

Identifying mRNA Editing Deaminase Targets by RNA-Seq

Brad R. Rosenberg, Scott Dewell, and F. Nina Papavasiliou

Abstract

RNA editing deaminases act on a variety of targets in different organisms. A number of such enzymes have been shown to act on mRNA, with the resultant nucleotide changes modifying a transcript's information content. Though the deaminase activity of mRNA editing enzymes is readily demonstrated *in vitro*, identifying their physiological targets has proved challenging. Recent advances in ultra high-throughput sequencing technologies have allowed for whole transcriptome sequencing and expression profiling (RNA-Seq). We have developed a system to identify novel mRNA editing deamination targets based on comparative analysis of RNA-Seq data. The efficacy and utility of this approach is demonstrated for APOBEC1, a cytidine deaminase with a known and well-characterized mRNA editing target in the mammalian small intestine.

Key words: RNA editing, Cytidine deaminase, Adenosine deaminase, APOBEC1, RNA-Seq

1. Introduction

RNA editing deaminases act on diverse targets in many organisms, ranging from bacteria to mammals. Editing enzymes that act on mRNA posttranscriptionally introduce considerable diversity to the transcriptome. Single nucleotide alterations can be sufficient to alter protein products (1) or modulate gene expression (2). Though the RNA editing activity of many deaminases is readily demonstrated *in vitro*, identifying physiological targets *de novo* has proved a challenge. As mRNA editing often modifies an individual base, comprehensively screening for new targets effectively means detecting single nucleotide changes across the entire transcriptome.

Recent advances in DNA sequencing technology provide new platforms for studying mRNA editing. Ultra high-throughput short read sequencing now allows for whole eukaryotic transcriptome sequencing and expression profiling (RNA-Seq) (3, 4). Using a similar approach, we have developed a method to identify

Ruslan Aphasizhev (ed.), *RNA and DNA Editing: Methods and Protocols*, Methods in Molecular Biology, vol. 718, DOI 10.1007/978-1-61779-018-8_6, © Springer Science+Business Media, LLC 2011

103

novel mRNA editing targets based on the comparative analysis of whole transcriptome sequencing data.

Following generation of a transcriptome-wide short read data set, the analysis workflow is geared specifically to identify mRNA editing events. As a comparative analysis technique, this approach requires mRNA isolated from cells/tissues/animals deficient in the particular deaminase under investigation, as well as from an editing-competent control. In the examples provided here, whole intestine tissue from *Apobec1*$^{-/-}$ mice and corresponding C57/Bl6 control animals are used to demonstrate proof-of-concept. APOBEC1 deaminates a specific cytidine to uridine in its well-characterized target transcript, apoB; editing is completely ablated in the knockout animal (5). If a genetic deaminase-null model system is impractical or unavailable, mRNA derived from other comparative systems (e.g., RNAi knockdown of a deaminase, overexpression in cell line systems, etc.) is also acceptable, provided the genetic backgrounds are suitably controlled. If an editing-deficient experimental system is not available, an alternative method for detecting mRNA editing has been described (6), though it does not provide a comprehensive whole transcriptome profile and cannot establish enzyme specificity.

The experimental strategy and analysis scheme are outlined in Fig. 1. Sample preparation and ultra high-throughput sequencing are similar to RNA-Seq procedures described elsewhere (3). Beginning with total RNA from control and deaminase-deficient cells, polyA⁺ mRNA is isolated and used to prepare RNA-Seq libraries. These are subjected to ultra high-throughput sequencing at high coverage. Sequencing reads from both samples are mapped independently to a reference genome, thereby generating separate "consensus" alignments. For each consensus, single nucleotide mismatches to the reference genome are identified using a "quality conscious" algorithm. These mismatch datasets are then filtered against several criteria, maintaining only those mismatches representative of potential deaminase-specific mRNA editing events. The two datasets are then intersected, with read:reference mismatches present in control mRNA sequences but absent from deaminase-deficient mRNA sequences output as potential editing hits. Finally, with genome and transcript coordinates, standard Sanger sequencing may be used to validate the novel mRNA editing targets.

2. Materials

2.1. RNA-Seq Library Preparation

1. MicroPoly(A)Purist Kit (Ambion, Austin, TX)
2. RNA fragmentation buffer (5×): 200 mM Tris–acetate, pH 8.2, 500 mM potassium acetate, 150 mM magnesium acetate
3. SuperScript III Reverse Transcriptase (Invitrogen, Carlsbad, CA)

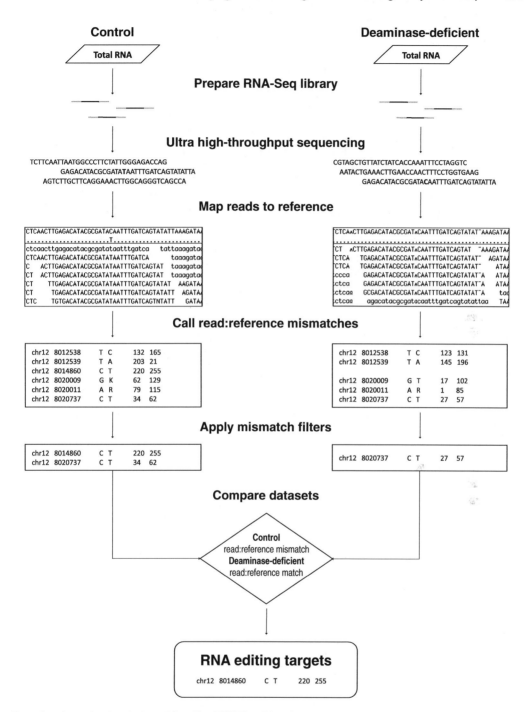

Fig. 1. Experimental and analysis workflow. The APOBEC1 editing site on the apoB transcript is highlighted as an example.

4. Random Primers (Invitrogen, Carlsbad, CA)

5. SuperScript Double-Stranded cDNA Synthesis Kit (Invitrogen, Carlsbad, CA)

6. QIAquick PCR Purification Kit (Qiagen, Valencia, CA)

7. Klenow Fragment $(3' \rightarrow 5'$ exo-$)$ (New England Biolabs, Ipswich, MA)

8. Quick Ligation Kit (New England Biolabs, Ipswich, MA)

9. PE Adapter Oligo Mix (Illumina, San Diego, CA)

10. HyLadder 1 kb (Denville Scientific, Metuchen, NJ)

11. QIAquick Gel Extraction Kit (Qiagen, Valencia, CA)

12. Phusion High-Fidelity DNA Polymerase (New England Biolabs, Ipswich, MA)

13. PCR Primers PE 1.0 and PE 2.0 (Illumina, San Diego, CA)

14. Nanodrop 1000 spectrophotometer (Thermo Scientific, Wilmington, DE)

15. *Optional* Agilent 2100 Bioanalyzer (plus RNA 6000 Pico kit) (Agilent Technologies, Santa Clara, CA)

2.2. Ultra High-Throughput Sequencing

1. Single-Read Cluster Generation Kit v2 (Illumina, San Diego, CA)

2. 36-Cycle Illumina Sequencing Kit v2 (Illumina, San Diego, CA)

3. Illumina Genome Analyzer II (Illumina, San Diego, CA)

4. GA Pipeline 1.4.0 software (Illumina, San Diego, CA)

2.3. Software and Computer Requirements

2.3.1. Recommended Configuration for Mapping and Analysis Computer

1. Short read mapping and alignment software for mapping RNA-Seq reads to reference genome: FASTX Toolkit (available at http://hannonlab.cshl.edu/fastx_toolkit/); Bowtie (available at http://bowtie-bio.sourceforge.net/index.shtml); TopHat (available at http://tophat.cbcb.umd.edu/)

2. Consensus and mismatch calling software for read:reference mismatch analysis: SAMTools (available at http://samtools.sourceforge.net/)

3. Recommended configuration for mapping and analysis computer: 2.8 GHz Intel Core 2 Duo or equivalent (additional cores will improve mapping and alignment performance); 8 GB RAM; Linux OS (any release)

3. Methods

The methods described in Subheadings 3.1–3.5 are performed in parallel for both control and deaminase-deficient samples.

3.1. RNA-Seq Library Preparation

The following RNA-Seq library preparation protocol was adapted from Ref. (3) and Illumina product literature.

1. Total RNA can be isolated from cells/tissue by standard methods of choice (see Note 1). The starting total RNA should be of high quality as determined by gel electrophoresis

or Agilent Bioanalyzer 2100 analysis (Fig. 2). For a typical preparation, starting with more than 25 µg of total RNA is recommended, though we have had success with less than 10 µg total RNA (see Note 2).

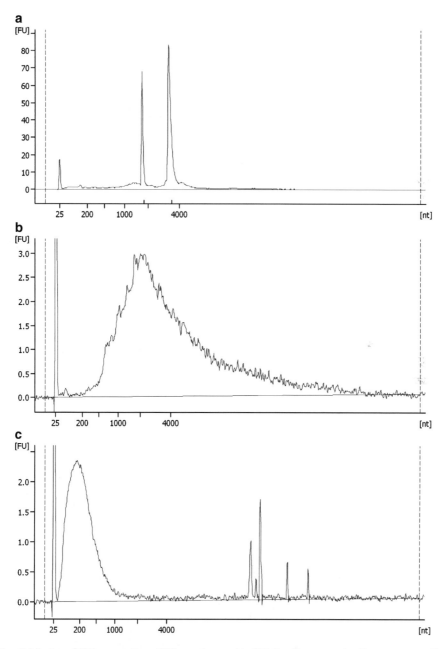

Fig. 2. Size distribution of RNA preparations. RNA samples used in RNA-Seq library construction were run on the Agilent 2100 Bioanalyzer. (**a**) Total RNA. Two sharp peaks corresponding to 18s and 28s rRNA are clearly visible. (**b**) PolyA⁺ mRNA. The broad range of transcript lengths is apparent and no residual small RNA or rRNA peaks are visible. (**c**) Fragmented polyA⁺ mRNA. Though somewhat variable in size, fragmented RNA is distributed about a peak length of approximately 200 nt, appropriate for RNA-Seq.

2. PolyA$^+$ mRNA is isolated from total RNA using the MicroPoly(A)Purist Kit. RNA is incubated on the polyT resin at room temperature for 1 h to ensure maximum binding and recovery of polyA$^+$ mRNA.

3. Following initial polyA$^+$ mRNA enrichment, step 2 is repeated with fresh polyT resin to maximize depletion of non-mRNA species (see Note 3). Purity and size distribution of enriched polyA$^+$ mRNA can be monitored by Agilent Bioanalyzer 2100 (Fig. 2).

4. The polyA$^+$ mRNA is concentrated by standard ethanol precipitation (plus glycogen) with a 70% ethanol wash. The pellet is resuspended in an appropriate volume of RNAse-free water to 100 ng/μl by OD$_{260}$.

5. 2 μl RNA fragmentation buffer (5×) is added to 8 μl polyA$^+$ mRNA on ice. The mRNA is fragmented by incubation at 94°C for exactly 4 min 30 s (see Note 4). The tubes should be transferred immediately to ice.

6. Fragmented mRNA is concentrated by standard ethanol precipitation (plus glycogen) with a 70% ethanol wash. The RNA pellet is resuspended in 14 μl water.

7. The distribution of mRNA fragment sizes can be evaluated by the Agilent Bioanalyzer 2100 (see Fig. 2).

8. First-strand cDNA is prepared using Superscript III Reverse Transcriptase (see Note 5) with random primers (150 ng/μl in 20 μl reaction volume). The reverse transcription reaction is incubated at 51°C for 45 min prior to enzyme inactivation at 70°C.

9. Second-strand synthesis is performed using the SuperScript Double-Stranded cDNA Synthesis Kit. Following the 2 h synthesis reaction, T4 DNA polymerase is added to fill-in ends.

10. The double-stranded cDNA (ds-cDNA) is purified and concentrated using the QIAquick PCR Purification Kit. Product is eluted in 30 μl water.

11. Adenine overhangs are added to the ds-cDNA by Klenow Fragment (3′→5′ exo-) in the presence of 200 mM dATP. The reaction is incubated at 37°C for 30 min.

12. The ds-cDNA (with A overhangs) is purified and concentrated using the QIAquick PCR Purification Kit. The product is eluted in 30 μl water.

13. Illumina sequencing adaptors are ligated to the ds-cDNA duplexes using the Quick Ligation kit. The reaction should contain 28 μl ds-cDNA, 30 μl 2× Quick Ligase buffer, 1 μl of the Illumina PE adaptor oligo mix, and 1 μl of T4 DNA Quick Ligase. The adaptors are ligated at room temperature for 15 min.

14. The ds-cDNA (with sequencing adaptors) is concentrated using the QIAquick PCR Purification Kit. The product is eluted in 30 μl of water.

15. The adaptor-ligated ds-cDNA samples (mixed with DNA loading buffer) are run on a 1% agarose-TAE gel (see Note 6) containing ethidium bromide (or comparable nucleic acid stain). Alternating lanes should be loaded with 15 μl 1 kb HyLadder, such that each sample lane is bordered on both sides by a size marker lane. The gel is run in TAE buffer at 100 V for 1 h.

16. The DNA is visualized on a UV transilluminator. Due to the small amount of cDNA, the sample lanes are typically not visible. As such, the alternating size marker lanes serve as a guide to extract the cDNA duplex size of interest. For each sample lane, a gel slice equivalent to 200±25 nt should be excised with a clean razor blade or scalpel and transferred to a clean microfuge tube.

17. The adaptor-ligated ds-cDNA is extracted and purified with the Qiagen Gel Extraction kit. Isopropanol must be added to the gel extraction buffer, as it is essential for efficient recovery of the short cDNA duplexes (~200 nt). The purified product is eluted in 30 μl water.

18. The RNA-Seq library is prepared by amplifying the adaptor-ligated ds-cDNA with Phusion high-fidelity DNA polymerase. The amplification reaction should contain 10 μl of 5× Phusion buffer, 1 μl of PCR Primer PE 1.0, 1 μl of PCR Primer PE 2.0, 1 μl of dNTPs (10 mM), 0.5 μl Phusion DNA polymerase, 6 μl water, and 30 μl of the ds-cDNA gel extraction product. Thermal cycler conditions: $98°C \times 10$ s, $65°C \times 30$ s, $72°C \times 30$ s for 15 cycles (see Note 2).

19. The amplified RNA-Seq library is purified using the QIAquick PCR Purification Kit. The product is eluted in 30 μl water.

20. The concentration of the RNA-Seq library is determined by Nanodrop spectrophotometer.

21. If desired, the size-based gel extraction and high-fidelity amplification steps can be checked by agarose gel electrophoresis. Approximately 5 μl of cDNA sequencing library (in DNA loading buffer) is loaded to a 1% agarose–TAE gel containing ethidium bromide (or comparable nucleic acid stain). One gel lane is reserved for DNA size markers (1 kb HyLadder). The gel is run in TAE buffer at 100 V for 1 h and the DNA is visualized by UV transillumination. A successful prep should display a broad band centered at approximately 250 bp.

3.2. Ultra High-Throughput Sequencing

Ultra high-throughput sequencing on the Illumina Genome Analyzer II (GAII) is detailed in the corresponding Illumina technical literature (7, 8). A brief summary of the technique as it applies to this protocol appears here.

1. The ds-cDNA RNA-Seq library is diluted to an appropriate concentration, typically 4–8 pM (see Note 7).

2. The RNA-Seq library is hybridized to a GAII flowcell, on which covalently linked oligomers (complimentary to the library adapter sequences) capture the ds-cDNA templates.

3. Template bridge amplification is performed at 55°C, in which hybridized templates are replicated onto the flowcell-bound oligomers.

4. Original library templates are removed by denaturing and washing.

5. Bridged templates are polymerase amplified, thereby generating flowcell-bound, sequence-matched "clusters."

6. Clusters are "cleaned up" to standardize strand polarity, remove unbound oligomers, and block nonclustered 3' ends (see Note 8).

7. Ultra high-throughput sequencing data is acquired by standard sequencing-by-synthesis reaction (see Note 9). This is typically performed in parallel for eight flowcell lanes. In the present example, acquisition is cycled for 36 nt reads (see Note 10). For RNA-Seq, one lane should be reserved for the Illumina-provided phiX174 library control.

8. Raw image data are processed using the standard Illumina software pipeline (SCS2.4). Real-time analysis and base calling generates files containing data on each sequencing read (*qseq.txt), intensities (*.cif) and noise profiles (*.cnf).

9. Spectral crosstalk and phasing (see Note 11) should be corrected using the Bustard (GA Pipeline 1.4.0) program and the phiX174 control lane as follows:

```
$ bustard.py --CIF --matrix=auto4 --phasing=auto4
Data/Intensities --make
$ nohup make -j N &
```

where N is the number of processors for analysis.

10. The resulting *qseq.txt files contain read IDs, sequence, and quality scores for each flowcell "tile." Standard FASTQ files are generated using the Gerald (GA Pipeline 1.4.0) program.

3.3. Mapping RNA-Seq Reads to Reference Genome

1. Trim 5' bases of RNA-Seq reads

Empirical evidence suggests that priming reverse transcription with random primers leads to an overrepresentation of G and C residues in the initial sequencing cycles. To eliminate this potential source of error, prior to mapping, each short read is "trimmed" – the 5' first two bases (and associated quality information) are digitally removed from the FASTQ data files. Such trimming can be accomplished through a variety of text edit script commands. We have had success with a simple and publicly available solution: the FASTQ/A Trimmer tool, part of the FASTX Toolkit software package.

General syntax:

```
$ fastx_trimmer [-f First base to keep] [-i INFILE]
[-o OUTFILE]
```

Example (trim 5′ first 2 bases):

```
$ fastx_trimmer -f 3 -i C57Bl6_Int.fq -o
C57Bl6_Int.trim2.fq
```

2. Map RNA-Seq reads to reference genome

Mapping RNA-Seq reads to a reference genome presents a number of computational challenges. Sequencing libraries derived from randomly fragmented mature mRNAs will generate a significant proportion of reads spanning exon–exon junctions. At the time of writing, there are several academic and commercial software packages available for aligning RNA-Seq short read data. Many are compatible with this workflow, though alignment parameters should be adjusted to reflect the demands of detecting nucleotide variations in mRNA libraries:

(a) Alignments should allow for mismatches in reads relative to the reference genome

(b) Alignments should be "quality conscious." Due to the relatively high error rate of ultra high-throughput sequencing and relatively low probability of an mRNA editing event, base quality scores should be taken into account for mapping and mismatch calling algorithms.

(c) Alignments should be unique. As an mRNA editing event is detected as a read mismatch to reference, it is imperative that mismatches occur only at "real" editing sites and not as a result of sequencing errors. Therefore, the mapping algorithm should suppress all reads for which an alignment is not "unique," i.e., it can be satisfactorily mapped to more than one location in the genome while still satisfying mismatch and quality limits. Though such an approach will dramatically reduce the number of potentially "good" alignments, such stringency is recommended to ensure high confidence in mismatch hits.

(d) Reads spanning exon–exon junctions should be mapped accordingly. Different alignment algorithms approach the problem of mapping reads derived from mature, spliced mRNAs to genomic reference with various strategies; these include alignment to an artificial "splice-ome" reference sequence of all predicted exon–exon junctions, and *ab initio* mapping of reads to junctions predicted by read distribution and pileup. Most strategies are appropriate for mRNA editing studies. Most importantly, mapping results should be written to a single comprehensive output consisting of both exonic and splice junction alignments; separate outputs for these alignments can lead to difficulties in mismatch analysis.

Taking the above criteria into account, we use the "Tuxedo Tools," TopHat, (9) and Bowtie (10) for mapping RNA-Seq reads. A command line example with appropriate parameters set appears below.

Example (mapping 36 nt raw reads trimmed to 34 nt):

```
$tophat -n 2 -g 1 -a 12 -m 1 --segment-length
34 mm9_genome C57Bl6_Intestine.trim2.fq
```

– n 2 allow for up to 2 mismatches to reference in seed region (first 28 nt); quality conscious

– g 1 suppress all alignments for reads that map to >1 location in reference

– a 12 for exon–exon junction reads, require at least 12 bases ("anchor") on either side of the junction

– m 1 for exon–exon junction reads, allow for up to one mismatch in anchor segment

– segment-length 34 read length is 34 nt (do not split reads)

TopHat alignments are output to a single comprehensive SAM file, which can be used in downstream analyses.

3.4. Call Read: Reference Mismatches

Once suitable RNA-Seq alignments are generated, single nucleotide mismatches to the reference sequence are identified. There are few (if any) software options specifically designed to call mismatches generated by RNA editing. However, several analysis packages incorporate SNP calling algorithms, which can often be implemented for RNA mismatch analysis. Most importantly, as for the alignment step, mismatch calling should be "quality conscious" to ensure high confidence in read:reference discrepancies.

Beginning with a TopHat generated SAM file, we use the publicly available, open source SAMTools software package (11) from the Sanger Institute for mismatch calling. The complete workflow can be found at http://samtools.sourceforge.net/samtools.shtml; an abridged example appears below.

Example of SAMTools workflow

1. TopHat output SAM file to SAMTools Binary (BAM) conversion

```
$./samtools view -S -b -t mm9_genome.fa.fai
C57Bl6_Intestine.accepted_hits.sam -o C57Bl6_
Intestine.bam
```

2. BAM file sort

```
$./samtools sort C57Bl6_Intestine.bam C57Bl6_
Intestine.sorted
```

3. BAM file indexing for rapid lookup functions

```
$./samtools index C57Bl6_Intestine.sorted.bam
```

4. Pileup conversion and consensus calling

```
$./samtools pileup -f mm9_genome.fa -c C57Bl6_
Intestine.sorted.bam > C57Bl6_Intestine.CNS
```

Generates a "pileup" file relative to the reference genome. This file contains reads and their qualities on a reference base-by-base scale and a corresponding consensus base call at each position.

5. Mismatch calling and preliminary filter

```
$./samtools.pl varFilter -d 3 -D 1000 -Q 25
C57Bl6_Intestine.CNS > C57Bl6_Intestine.fil-
tered.MMtoRef
```

Filters the pileup consensus file for variations in reads relative to reference. This dataset will serve as the starting point for identifying those mismatches that resulted from mRNA editing. In this example, a relatively stringent filter is applied requiring a minimum of three mismatch-containing reads (-d 3) to call a variant. The maximum parameter (-D 1000) should be adjusted based on the read depth achieved in a given experiment. Finally, the root mean squared mapping quality must exceed 25 (-Q 25) to call a mismatch. Additional parameters can be adjusted as needed.

3.5. Read:Reference Mismatch Filters

The list of variations called from consensus will often contain large numbers of read:reference mismatches unrelated to mRNA editing. These mismatches may be a result of genomic SNPs, sequencing errors, misaligned reads, reverse transcription/amplification errors, and unrelated mRNA modification processes. As such, the initial variation list is filtered on several criteria appropriate to the editing enzyme under investigation, including error probability, sequence type (exons of known or predicted genes only), read:reference mismatch base calls, known SNPs, and repetitive elements. Filters can be applied by many different bioinformatics packages and/or standard Linux shell commands. We use Galaxy, a web-based genomics suite (http://galaxy.psu.edu/), largely because of its user-friendly interface and powerful genomic interval operations. A particular series of filters will be unique to each mRNA editing enzyme studied. A sample filter workflow for an mRNA-editing cytidine deaminase appears below. This example uses a particularly stringent set of filters to minimize false-positive hits; parameters should be adjusted to suit a given experiment.

1. Error probability filter

   Keep only those mismatches with a PHRED-scaled probability score ≥40 (as calculated by SAMtools pileup function).

2. Sequence type filter

   Keep only those mismatches within mRNA exons (from RefSeq database, http://genome.ucsc.edu/cgi-bin/hgTables).

3. Base call filter

   Keep only C:T (reference:read) mismatches for + strand transcripts (G:A mismatches for – strand transcripts) (see Note 12).

4. Known SNP filter

   Keep only those mismatches absent from SNP databases (see Note 13).

5. Repetitive element filter

   Keep only those mismatches outside of repetitive regions (see Note 14).

   Subsequent analysis steps are performed on a single, merged dataset derived from the above control and deaminase-deficient mismatch results.

3.6. Identifying RNA Editing Targets

1. Mismatches specific to deaminase activity are identified by intersecting the control and deaminase-deficient analyses. Read:reference mismatches present in the control dataset and absent from the deaminase-deficient dataset are flagged as potential mRNA editing hits.

2. Potential editing events are further refined by comparing read consensus between control and deaminase-deficient datasets at each mismatch hit site. Consensus pileup data (Subheading 3.4, step 4) from each library are juxtaposed at each mismatch hit site. Potential hits in the control dataset should have a multiple read, reference-matching consensus with low error probability at the corresponding site in the deaminase-deficient dataset. Using SAMtools, we define this as a minimum read-depth ≥3 with a PHRED-scaled probability score ≥30. Mismatches that do not satisfy these criteria are discarded.

3. Potential editing hits are tabulated and organized with associated consensus data and error probabilities for both datasets. As hits are reported with coordinates on the genomic reference, the results can be additionally annotated with information from a wide variety of genome resource databases. Pileup data can also be used to calculate "editing frequency," as determined by the percentage of reads representing the edited base in the control dataset. Results for APOBEC1 in mouse small intestine appear in Table 1.

Table 1
APOBEC1 mRNA editing targets

| Site annotation | | | | Reference base | C57/Bl6 | | | | | APOBEC1[−/−] | | | |
Genome coordinate	Transcript ID	Gene name	Strand		Read consensus	Consensus error probability	Mismatch error probability	Read depth	Editing frequency (%)	Read consensus	Consensus error probability	Mismatch error probability	Read depth
chr12:8014860	NM_009693	ApoB	+	C	T	255	255	266	94.4	C	255	0	200
chrX:109671648	NM_026333	2010106E10Rik	+	C	Y	228	228	181	43.1	C	254	0	99
chr8:46391931	NM_133969	Cyp4v3	−	G	R	228	228	105	32.4	G	129	0	34
chr2:121978638	NM_009735	B2m	+	C	Y	228	228	323	26.0	C	255	0	173
chr16:84955113	NM_007471	App	−	G	R	228	228	82	28.0	G	250	0	74
chrX:136207009	NM_023270	Rnf128	+	C	Y	216	216	167	22.2	C	219	0	87

RNA-Seq libraries from the small intestine of control C57/Bl6 and *Apobec1−/−* mice were prepared and screened for mRNA editing events. Sorted by PHRED-scaled mismatch probability score, the "top hit" is a C–T change at position 6666 of the apoB transcript, the well-characterized physiological substrate of APOBEC1. Additional, previously unknown APOBEC1 editing targets were also identified; a partial list of these hits is provided here

3.7. Validating Potential RNA Editing Targets

Newly identified mRNA editing targets should be validated by standard Sanger sequencing.

1. Regions of interest (containing editing hits) are amplified by high-fidelity PCR. Identical regions are amplified from control genomic DNA, control cDNA, deaminase-deficient genomic DNA, and deaminase-deficient cDNA.

2. PCR products are sequenced by standard Sanger methods.

3. mRNA editing hits are validated by comparing the four sets of Sanger sequencing results. A *bona fide* mRNA editing hit is confirmed by reference-matching sequence in control genomic DNA, deaminase-deficient genomic DNA, and deaminase-deficient cDNA, but reference-mismatch sequence in control cDNA (see Note 15). An example is shown in Fig. 3.

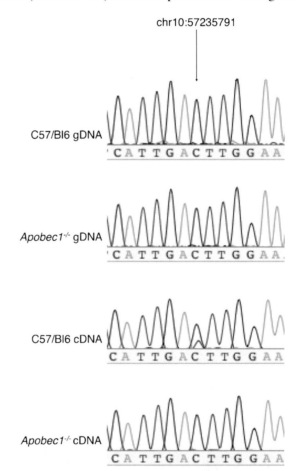

Fig. 3. mRNA editing target validation by Sanger sequencing. Regions spanning a potential editing site (chr8:46391931, NM_133969, 32.4% editing frequency by RNA-Seq) were PCR amplified from genomic DNA and cDNA isolated from control C57/Bl6 and *Apobec1⁻/⁻* mouse small intestine. PCR products were sequenced by standard Sanger methods. The C:T modification characteristic of APOBEC1 editing is present only in the control C57/Bl6 cDNA.

4. Notes

1. Total RNA isolated by most standard molecular biology techniques is appropriate for prepping RNA-Seq libraries. However, certain spin column-based total RNA kits may limit the recovery of very long and/or very short transcripts. Total RNA isolated with such kits remains acceptable for library prep, but may bias against certain transcript types.

2. If less than 10 µg total RNA is used, additional PCR amplification cycles may be added at step 18 (Subheading 3.1). We have had success amplifying 15 cycles for >10 µg starting total RNA, 16 cycles for 5–10 µg starting total RNA and 17 cycles for 4 µg starting total RNA. Though additional amplification cycles can provide sufficient library concentration from limited starting material, it also increases polymerase errors and may bias the RNA-Seq dataset.

3. Though the final yield of polyA⁺ mRNA is somewhat reduced, a second round of polyA⁺ enrichment is necessary to completely remove rRNA, tRNA, and other species that interfere with RNA-Seq.

4. The ideal mRNA fragment size for a standard RNA-Seq run on the Illumina GA II is 200 nt. If required for alternative applications, the fragment size can be adjusted by altering the duration of the reaction. Longer fragments can be achieved by reducing incubation time at 94°C and vice versa.

5. Though the SuperScript Double-Stranded cDNA Synthesis Kit includes a reverse transcriptase (Superscript II), this protocol substitutes a different enzyme (Superscript III). Superscript III can be used with the same buffers described in the ds-cDNA synthesis kit but functions at a higher temperature (51°C), thereby minimizing potential secondary structure in RNA fragments.

6. RNA-Seq ultra high-throughput sequencing is most effective with cDNA fragments of relatively short (~200 nt) and defined size. As fragmentation can be variable and somewhat heterogeneous, ds-cDNA libraries are separated by agarose gel electrophoresis and extracted based on DNA size standards.

7. An appropriate concentration range allows each ds-cDNA template to be spatially separated on the flowcell at a distance sufficient for cluster amplification and resolution while maximizing total template number.

8. The clean up steps remove all extended nucleotides from one of the flowcell-bound oligomers, leaving only the sequence extended from the other oligomer. This guarantees that only one of the strands will be sequenced during acquisition.

9. The sequencing reaction consists of stepwise cycles that proceed iteratively; total cycle number dictates read length. Polymerase extends the 3′ end of primer (annealed in the final amplification step) through the incorporation of reversibly terminating nucleotides (fluorescently labeled by base). After the incorporation of one chain-terminating nucleotide, the entire flowcell is imaged in small "tiles," 120 tiles per lane. Four images are captured per tile, one for each nucleotide laser/ filter combination. A cleavage step removes the fluorophore and reverses chain termination. After a wash step, the cycle is repeated until the desired read length has been reached.

10. RNA-Seq read depth required for mismatch/editing analysis will vary based on the size of the genome/transcriptome under investigation and the read length. For mammalian transcriptomes (mouse, human), at 36 nt read lengths, approximately 50–60 million raw unmapped reads should provide comprehensive transcriptome coverage (>95% of all bases in expressed transcripts).

11. Due to the nature of the sequencing methodology, certain calculations and corrective measures are implemented to account for spectral crosstalk from the fluorophores and the fact that not all strands in each cluster are extended in perfect synchrony; some will not be extended and others will be extended by two bases at certain cycles, leading to what is termed "phasing." For RNA-Seq it is important to include a sequencing control sample in one lane of the flowcell; the phiX174 library provided is appropriate for this purpose.

12. As RNA-Seq reads are mapped to a reference genome, all mismatch bases and coordinates are provided in the context of the genomic forward strand. Therefore, for an analysis of mRNA editing, transcript "strandedness" must be taken into account. When filtering for cytidine deaminase editing, both C–T mismatches for + strand transcripts AND G–A mismatches for – strand transcripts pass. Similarly, for adenosine deaminases, A–G mismatches for + strand transcripts AND T–C mismatches for – strand transcripts pass.

13. Genomic heterozygosity and SNPs are detected as read:reference mismatches and can lead to false positive mRNA editing hits. To minimize this error source, known SNPs are removed from the dataset. Known SNP databases for a variety of genomes are available from UCSC: http:// genome.ucsc.edu/cgi-bin/hgTables.

14. Repetitive element databases (such as RepeatMasker) for a variety of genomes are available from UCSC: http://genome. ucsc.edu/cgi-bin/hgTables.

15. Hit validation requires standard Sanger sequencing of all four samples to rule out several potential non-mRNA editing sources of read:reference mismatch. First, genomic DNA from each library is sequenced to rule out heterozygosity. Next, genomic DNA sequences from control and deaminase-deficient samples are compared to rule out a SNP. Control cDNA sequence is compared to control genomic DNA sequence to confirm mRNA editing. Finally, deaminase-deficient cDNA sequence is compared to control cDNA sequence to validate specific deaminase activity.

References

1. Wedekind, J. E., Dance, G. S., Sowden, M. P., and Smith, H. C. (2003) Messenger RNA editing in mammals: new members of the APOBEC family seeking roles in the family business, *Trends Genet 19*, 207–216.

2. Nishikura, K. (2006) Editor meets silencer: crosstalk between RNA editing and RNA interference, *Nat Rev Mol Cell Biol 7*, 919–931.

3. Mortazavi, A., Williams, B. A., McCue, K., Schaeffer, L., and Wold, B. (2008) Mapping and quantifying mammalian transcriptomes by RNA-Seq, *Nat Methods 5*, 621–628.

4. Nagalakshmi, U., Wang, Z., Waern, K., Shou, C., Raha, D., Gerstein, M., and Snyder, M. (2008) The transcriptional landscape of the yeast genome defined by RNA sequencing, *Science 320*, 1344–1349.

5. Hirano, K., Young, S. G., Farese, R. V., Ng, J., Sande, E., Warburton, C., Powell-Braxton, L. M., and Davidson, N. O. (1996) Targeted disruption of the mouse apobec-1 gene abolishes apolipoprotein B mRNA editing and eliminates apolipoprotein B48, *J Biol Chem 271*, 9887–9890.

6. Li, J. B., Levanon, E. Y., Yoon, J. K., Aach, J., Xie, B., Leproust, E., Zhang, K., Gao, Y., and Church, G. M. (2009) Genome-wide identification of human RNA editing sites by parallel DNA capturing and sequencing, *Science 324*, 1210–1213.

7. Illumina. (2008) Using the Single-Read Cluster Generation Kit v2 on the Cluster Station, San Diego, CA.

8. Illumina. (2008) Using SBS Sequencing Kit v3 on the Genome Analyzer, San Diego, CA.

9. Trapnell, C., Pachter, L., and Salzberg, S. L. (2009) TopHat: discovering splice junctions with RNA-Seq, *Bioinformatics 25*, 1105–1111.

10. Langmead, B., Trapnell, C., Pop, M., and Salzberg, S. L. (2009) Ultrafast and memory-efficient alignment of short DNA sequences to the human genome, *Genome Biol 10*, R25.

11. Li, H., Handsaker, B., Wysoker, A., Fennell, T., Ruan, J., Homer, N., Marth, G., Abecasis, G., Durbin, R., and Subgroup GPDP. (2009) The Sequence Alignment/Map format and SAMtools, *Bioinformatics 25*, 2078–2079.

Chapter 7

Mouse and Other Rodent Models of C to U RNA Editing

Valerie Blanc and Nicholas O. Davidson

Abstract

Substitutional RNA editing represents an important posttranscriptional enzymatic pathway for increasing genetic plasticity by permitting production of different translation products from a single genomically encoded template. One of the best-characterized examples in mammals is C to U deamination of the nuclear apolipoprotein B (apoB) mRNA. ApoB mRNA undergoes a single, site-specific cytidine deamination event yielding an edited transcript that results in tissue-specific translation of two distinct isoforms, referred to as apoB100 and apoB48. Tissue- and site-specific cytidine deamination of apoB mRNA is mediated by an incompletely characterized holoenzyme containing a minimal core complex consisting of an RNA-specific cytidine deaminase, Apobec-1 and a requisite cofactor, apobec-1 complementation factor (ACF). The underlying biochemical and genetic mechanisms regulating tissue-specific apoB mRNA editing have been accelerated through development and characterization of physiological rodent models as well as knockout and transgenic animal strains.

Key words: Lipid metabolism, RNA editing, Apobec-1, Hepatocytes, Hormonal regulation, Diet, Primer extension, Subcellular distribution

1. Introduction

Substitutional cytidine to uridine (C to U) RNA editing is an enzymatic process that alters a genomically encoded sequence through site-specific deamination of cytidine residues in nuclear RNA transcripts. C to U RNA editing of the apolipoprotein B (apoB) mRNA has been extensively characterized and involves enzymatic deamination of a C to U base in the nuclear apoB mRNA that converts a glutamine codon (CAA) into a stop codon (UAA), leading to a premature interruption of translation and synthesis of a truncated form of apolipoprotein B, defined as apoB48. The unedited version of apoB mRNA encodes a full-length protein, apoB100 (1, 2).

Ruslan Aphasizhev (ed.), *RNA and DNA Editing: Methods and Protocols*, Methods in Molecular Biology, vol. 718,
DOI 10.1007/978-1-61779-018-8_7, © Springer Science+Business Media, LLC 2011

Posttranscriptional C to U deamination of apoB mRNA is regulated in a species and tissue-specific manner. ApoB mRNA editing, and consequently apoB48 synthesis, occurs in enterocytes of the small intestine of all mammals. For yet incompletely understood reasons some species, such as mice and rats, also exhibit apoB RNA editing in the liver (2) and secrete both apoB100 and apoB48. Human liver, by contrast, synthesizes and secretes only apoB100 (3).

Introduction of a translational stop codon in the edited apoB transcript leads to translation of a carboxy-terminal truncated protein (apoB48) that lacks key domains present in the full-length (apoB100) protein. ApoB100-containing lipoproteins are recognized by a ubiquitously expressed low-density lipoprotein receptor (LDLR) through interactions with a domain localized in the C terminus of apoB100. Because they lack the LDLR binding domain present in apoB100, apoB48-containing lipoproteins are cleared through a distinctive receptor expressed predominantly in the liver (4). Thus, intestinal apoB48-containing lipoproteins represent an evolutionary adaptation to efficiently deliver dietary triglycerides and fat-soluble vitamins to the liver. The fine regulation of plasma cholesterol homeostasis by contrast is regulated via LDLR-mediated uptake of apoB100-containing lipoproteins.

The enzymatic deamination of apoB mRNA is mediated by a holoenzyme whose minimal core is composed of an RNA-specific cytidine deaminase (Apobec-1) (5) and an essential RNA-binding protein cofactor, apobec-1 complementation factor (ACF) (6, 7). Apobec-1 expression is limited to the stomach, small intestine, and colon in humans, but is not expressed in human liver (3). By contrast, ACF shows a broad expression pattern both in rodents and humans, with greatest abundance in liver, small intestine, and kidney (6, 8).

Both apobec-1 and ACF mRNA and protein expression undergo developmentally regulated increase in the fetal and neonatal small intestine, presumably to accommodate some of the developmental changes taking place in lipid metabolism in which the relative production of either apoB48 or apoB100 would confer a metabolic advantage (9, 10). This prediction is consistent with the numerous studies demonstrating tissue-specific regulation of C to U RNA editing in the setting of metabolic adaptations involved in lipid metabolism. This review addresses the approaches and methods to studying developmental, nutritional, hormonal, and genetic modulation of C to U RNA editing in murine and other rodent models.

2. Materials

2.1. Animals

2.1.1. Congenitally Hypothyroid Mice (Pax8⁻/⁻)

$Pax8^{-/-}$ mice are generated by homologous recombination (11). $Pax8^{-/-}$ mice manifest a phenotype of early growth delay and death after weaning. These animals are characterized by the absence of thyroxine-producing follicular cells in the thyroid, which results in the inability to autonomously produce thyroid hormone at weaning (12). $Pax8^{-/-}$ mice exhibit decreased hepatic apoB RNA editing associated with reduced mRNA and protein levels of ACF (13).

2.1.2. Zucker Rats

Adult fatty Zucker rats (*fa/fa*) obtained from Harlan Industries (Indianapolis, IN) carry a missense (A to C) mutation in the leptin receptor (OB-receptor) (14). *fa/fa* rats are characterized by obesity, hypercholesterolemia, hyperlipidemia, hyperglycemia, and hyperinsulinemia. This rat model of acquired insulin resistance recapitulates the observations reported in isolated rat hepatocytes incubated with insulin (15). Specifically, Zucker rats demonstrate elevated hepatic apoB mRNA editing in conjunction with increased Apobec-1 mRNA abundance (16). Insulin may itself modulate ACF expression since incubation of rat primary hepatocytes with 10 nM insulin produced a 1.5-fold increase of apoB editing activity that correlated with increased ACF nuclear localization (17). Further studies demonstrated that insulin-dependent ACF nuclear import is related to an increase in ACF phosphorylation (18). These findings suggest a new regulatory mechanism controlling apoB RNA editing.

2.1.3. Ob/Ob Mice

Mutant *ob/ob* mice in a C57BL/6J background are obtained from Jackson Laboratory (Bar Harbor, ME). These mice gain weight rapidly, exhibit hyperphagia, hyperglycemia, glucose intolerance, and insulin resistance with elevated plasma insulin levels. *In vivo* studies report an increased ratio of apoB100/apoB48 without changes in apobec-1 expression (19).

2.1.4. db/db Mice

The *db* gene encodes for the leptin receptor, Ob-R. The spontaneous homozygous *db* mutation results in the truncation of the Ob-Rb spliced form leading to loss of signaling activity (20). The *db/db* mice in a C57BL6/J background are obtained from Jackson Laboratory (Bar Harbor, ME). The mutant animals exhibit severe obesity and hypoglycemia coupled with hyperinsulinemia (21). *db/db* mice, like *ob/ob* mice, exhibit increased serum apoB100/apoB48 ratios whose mechanism is incompletely understood (22).

2.1.5. Apobec-1⁻/⁻ Mice

Three independent lines of $apobec-1^{-/-}$ lines were generated by homologous recombination (23, 24). Homozygous mice appear healthy and fertile with no alterations in serum cholesterol or triglyceride concentration. *Apobec-1$^{-/-}$* mice lack apobec-1 protein, fail to mediate C-to-U RNA editing in any tissue, and consequently synthesize and secrete exclusively apoB100 (23, 24).

2.1.6. Liver-Specific Apobec-1 Transgenic Mice

A full-length rabbit apobec-1 cDNA was cloned downstream of a liver-specific promoter (pLiv11) (25). Four lines of apobec-1 transgenic animal were generated ranging from 3 to 17 copies of the gene. Transgenic overexpression of rabbit apobec-1 results in loss of specificity for the canonical cytidine 6666 and C to U editing at multiple other cytidines in apoB RNA, a phenomenon known as hyperediting (26). In addition other RNAs undergo C to U RNA editing including the tumor suppressor gene NAT1 which undergoes multiple C to U modifications resulting in reduced NAT1 expression. This loss-of-function of NAT1 leads to hepatic dysplasia and hepatocellular carcinoma (25).

2.1.7. Tetracycline-Dependent Liver-Specific Conditional Apobec-1 Transgenic Mice

Rabbit apobec-1 was cloned downstream of a tetracycline-mediated transactivator (tTA) response element and a line of transgenic mice generated and crossed into a liver-activating protein (LAP)-tTA transgenic mice which express tTA specifically in hepatocytes (27). Thus, rabbit apobec-1 can then be turned on and off by removing from or adding tetracycline to the drinking water (28) (see Note 1).

2.2. Hormonal Regulation of C to U RNA Editing

2.2.1. Thyroid Hormone

3,5,3′-triiodo-L-thyronine (Sigma) (T3) stock solutions are prepared in saline containing 10 mg/mL bovine serum albumin adjusted to pH 11. Aliquots (20 μL) are kept frozen at –20°C.

2.2.2. Insulin

Bovine insulin (Sigma, St Louis, MO) is added to the feeding medium of primary hepatocytes in a concentration range of 0–67 nM (400 ng/mL).

2.2.3. Estrogen

17α-Ethinyl estradiol and 17β-estradiol stock solutions (Sigma, St Louis, MO) are prepared in propylene glycol (1 mg/mL).
Rodent show diet is from ICN Biochemicals, Cleveland, OH.

2.3. Dietary Modulation of C to U RNA Editing

2.3.1. Diet-Induced Hypothyroidism

Propylthiouracil (2-thio-4-hydroxy-6-*n*-propylpyrimidine) (Sigma, St Louis, MO) is dissolved in H_2O at a concentration of 1 mg/mL and aliquots are kept at –20°C.

2.3.2 Ethanol

Control liquid diet: 180 Kcal/L protein; 350 Kcal/L fat; 470 Kcal/L carbohydrate.

Ethanol liquid diet: 180 Kcal/L protein; 350 Kcal/L fat; 115 Kcal/L carbohydrate; 355 Kcal/L ethanol (6.7% v/v) (Bio-Serv Inc., Frenchtown, NJ).

2.3.3. High Carbohydrate

A high carbohydrate (58.45% sucrose) fat-free diet (ICN Nutritional Biochemicals) is used to augment hepatic lipogenesis and to stimulate the accumulation of fat in the liver.

2.3.4. Fasting–Refeeding Protocol

Rats (Sprague–Dawley, Charles River Wilmongton, MA).
    Purina rodent chow diet (Purina Mills).

2.4. Buffers and Media

2.4.1. Primary Hepatocyte Isolation Culture Medium

1. Male Sprague–Dawley rats or C57BL/6 mice (8–12 weeks old).

2. Anesthesia: 87 mg/kg ketamine HCl (Fortdodge Animal Health), 13.4 mg/kg xylazine (Lloyd Laboratories) (intraperitoneal administration).

3. "All purpose sponge" (2 × 2 in.) (Kendall).

4. Sterile catheter (22G × 1 in.) (Terumo).

5. [4$^{1/2}$″ straight] Iris scissors (Miltex Instruments).

6. 10× liver perfusion buffer: 1.2 M NaCl; 60 mM KCl; 22 mM NaH$_2$PO$_4$; 40 mM Na$_2$HPO$_4$; 5.5 M glucose; 25 mM MgSO$_4$. Working perfusion solution is prepared by diluting 10× perfusion buffer with NaHCO$_3$ (1.7% w/v final concentration) and 0.2 mM EGTA; pH 7.4. Filter the solution through 0.22-μm filter (Techno Plastic Product) and add 1 mL of 100× penicillin/streptomycin (Mediatech Cellgro).

7. Liver Digest Medium supplemented with Collagenase-Dispase (Invitrogen, Gibco).

8. Pump settings (Minipuls 2 peristaltic pump, Gilson): 3–4 mL/min. Wash tubing with 70% ethanol for 5 min followed by 5 min of 1× PBS and 5 min of perfusion working solution.

9. 100-μm filter (BD Falcon).

10. Collagen I-coated dish (35 mm) (BD Biosciences).

11. Wash medium: William's medium E (Invitrogen) supplemented with 0.64 mM L-ornithine, 38 mM sodium bicarbonate, 10 mM HEPES, 10 mM dextrose, 100 mg/mL streptomycin and 100 IU/mL penicillin G, 50 nM triiodothyronine, 1 mg/mL bovine serum albumin, 5 mg/mL linoleic acid, 0.1 mM CuSO$_4$, 3 nM NaSeO$_3$, 50 pM ZnSO$_4$.

12. L15 Medium (Invitrogen).

13. Culture medium: wash medium supplemented with 5% FBS and 1× penicillin/streptomycin.

2.4.2. Protein Extraction

1. Tissue lysis buffer: 20 mM Tris pH 8.9; 1 mM sodium vanadate; 150 mM NaCl; 2 mM EDTA; 100 mM sodium fluoride; 5% glycerol; 50 mM β-glycerophosphate and protease inhibitors (Roche Maxi or Mini (for 50 and 10 mL lysis buffer, respectively)).

2. 10× detergent buffer: 10% Triton X-100; 1% SDS in lysis buffer.

3. Dignam A: 10 mM HEPES pH 7.9; 1.5 mM $MgCl_2$; 10 mM KCl.

4. Dignam B: 300 mM HEPES pH 7.9; 1.4 M KCl; 30 mM $MgCl_2$.

5. Dignam D: 20 mM HEPES pH 7.9; 0.1 M KCl; 20% glycerol.

6. Protein concentration is assessed using Bio-Rad Protein assay (BIO-RAD).

2.4.3. RNA Analysis

1. RNA extraction: TRIzol (Invitrogen) following manufacturer's protocol.

2. DNAse treament: DNA-free kit (Ambion).

3. First strand synthesis: High Capacity cDNA Reverse Transcription Kit (Applied Biosystems). The reactions are performed in a 2720 Thermo Cycler (Applied Biosystems).

4. Quantitative PCR: SyBR GreenER qPCR SuperMix (Invitrogen). Amplifications are performed with an ABI Prism 7000 instrument (Applied Biosystems).

2.4.4. Poisoned Primer Extension

1. apoB amplification: Taq DNA polymerase (Invitrogen).

2. PCR purification: QIAquick PCR purification kit (QIAGEN).

3. Primer extension: 10× annealing buffer: 400 mM Tris–HCl pH 7.5.

4. 200 mM $MgCl_2$; 500 mM NaCl.

5. Extension buffer: 10 mM DTT; 25 mM d(ATP/dCTP/dTTP); 0.8 mM ddGTP; 1.5U T7 DNA polymerase (USB).

6. PhosphorImager/ImageQuant (GE Healthcare).

2.4.5. 10× In Vitro Conversion Buffer

100 mM HEPES, pH 7.9, 1 M KCl, 1 mM EDTA, 2.5 mM DTT, 10% glycerol, 250 ngtRNA and 20 units RNAsin Rinonuclease Inhibitor.

2.4.6. Cellular Fractionation Buffers

1. Hypotonic buffer A: 10 mM HEPES, 10 mM KCl, 1.5 mM $MgCl_2$, 0.1 mM EDTA, 1 mM DTT, and protease inhibitors.

2. Buffer B: Buffer A adjusted to 300 mM HEPES, 1.4 mM KCl, 30 mM $MgCl_2$.

3. Buffer C: 20 mM HEPES, 500 mM KCl, 1.5 mM MgCl$_2$, 0.1 mM EDTA, 1 mM DTT, and protease inhibitors.

4. Dialysis buffer: 20 mM HEPES (PH 7.9), 0.1 M KCl, 0.2 mM EDTA, 0.5 mM DTT, 20% glycerol, 0.5 mM benzamidine, 0.5 mM PMSF.

2.4.7. Immunohisto-chemistry Solutions

1. Fixative solution: 10% formalin (Sigma).

2. Rehydration solution: citrate buffer (0.01 M pH 6).

3. Blocking solution: 5% normal goat serum/3% bovine serum albumin.

Biotinylated goat anti-rabbit IgG (Vectastain Elite, Vector Labs, Burlingame, CA).
Avidin-conjugated horseradish peroxidase (Vector Labs).

2.4.8. Fast Protein Liquid Chromatography

Chromatography buffer: 1 mM EDTA, 150 mM NaCl, and 10 mM sodium phosphate, pH 7.4.

2.4.9. Analysis of Serum Lipids and Lipoproteins Content

L-type TG H, L type TG-M-ColorA and L type TG-M – ColorB kits (Wako Diagnostics).
Cholesterol E kit (Wako Diagnostics).

2.5. Primers

2.5.1. Quantitative PCR Primers

Murine apobec-1: Forward 5′-ACCACACGGATCAGCGGAAA-3′ Reverse 5′-TCATGATCTGGATAGTCACACCG-3′; murine apoB: Forward 5′-CACTGCCGTGGCCAAAA-3′ Reverse 5′-GCTAG-AGAGTTGGTCTGAAAAATCCT-3′; murine ACF: Forward 5′-GATGAAAAAAGTCACAGAAGGAGTTG-3′ Reverse 5′-CA-AATCCCCGGTTTTTGGT-3′; murine GAPDH Forward 5′-TGTGTCCGTCGTGGATCTGA-3′; Reverse 5′-CCTGCTT-CACCACCTTCTTGA-3′.

2.5.2. Murine apoB mRNA Amplification

Forward: 5′-ATCTGACTGGGAGAGACAAGTAGC-3′; Reverse: 5′-ACGGATATGATACTGTTCATCAAGAA-3′.

2.5.3. Poisoned Primer Extension (Antisense Primer)

5′-CCTGTGCATCATAATTATCTCTAATATACTGATCA-3′.

3. Methods

3.1. Hormonal Regulation of C to U RNA Editing

3.1.1. Thyroid Hormone

Chemically induced hypothyroid animals (mice or rats) are injected intraperitoneally with 3,5,3′-triiodo-L-thyronine (T3) at a physiological dose (0.5 mg/100 g of body weight) to normalize thyroid hormone levels or at pharmacological doses (50 mg/100 g of body weight) for 7 days to induce hyperthyroidism (29). For particular studies, when a time course of apoB mRNA editing is required, animals may undergo a single intravenous bolus of T3 at 100 mg/100 g of body weight a minimum of 3 h before determining apoB100 synthesis (29).

Congenitally hypothyroid (*Pax8*$^{-/-}$) mice (2-week-old) are injected intraperitoneally with a combination of T4 (2.5 mg/kg) and T3 (0.25 mg/kg) in 100 μL PBS daily for 2–4 days (12). This combination of TH has been demonstrated to promote hyperthyroidism in mice (30).

3.1.2. Insulin

Bovine insulin is added to the feeding medium of primary hepatocytes in a concentration range of 0–67 nM (400 ng/mL). The culture medium, with or without insulin, is replaced daily for a period of time ranging from 6 h to 5 days.

3.1.3. Estrogen

17α-Ethinyl estradiol solutions are administered subcutaneously (3–5 mg/kg/day) to 200–280 g Sprague–Dawley male rats for 5 consecutive days between 8:30 and 10:30 a.m. The animals have access to standard rodent chow diet. Twelve hours before sacrifice, the animals are fasted to reduce fluctuations of plasma cholesterol. Male mice are administered 17β-estradiol at a dose of 3 mg/kg body weight/day for 5 consecutive days. The animals have access to standard rodent chow. The animals are fasted overnight and sacrificed on the seventh day (31).

3.2. Dietary Modulation of C to U RNA Editing

3.2.1. Diet-Induced Hypothyroidy

The animals (mice or rats) are fed with chow diet supplemented with 0.1% (w/w) propylthiouracil (2-thio-4-hydroxy-6-*n*-propylpyrimidine) for 21–28 days in order to induce hypothyroidism followed by assessment of plasma thyroid hormone levels (T4, T3). Induction of hypothyroidism results in abnormal lipid metabolism and also changes C to U RNA editing in the liver of experimental animals.

3.2.2. Ethanol

Male Wistar rats (200–250 g) are fed regular rat chow for a minimum of 10 days before starting the diet regimen and are then acclimated for another 7 days to receiving a liquid diet as their unique feeding source. The amount of ethanol is progressively increased as followed: first 3 days: 1/3 ethanol 2/3 control diet; fourth to seventh day: 2/3 ethanol 1/3 control diet; eighth day on: pure ethanol diet. The diet is provided for a period of 15, 20, 30 up to 40 days. Alcohol feeding stimulates hepatic lipogenesis and results in hepatic fat accumulation (32).

3.2.3. High Carbohydrate

A high carbohydrate (58.45% sucrose) fat free diet is used to augment hepatic lipogenesis and to stimulate the accumulation of fat in the liver. This diet is given ad libitum for a minimum of 7 days prior to study.

3.2.4. Fasting–Refeeding Protocol

Rats are acclimated to a 12 h light/12 h dark cycle and fed a Purina rodent chow diet for 10 days and then fasted for 48 h. Selected groups are studied at the end of the 48 h fasting period. Other groups of rats are refed a high sucrose-fat free diet (high carbohydrate diet described above) for 24 or 48 h. This regimen induces dramatic hepatic lipid accumulation in the refed animals and results in alterations in C to U RNA editing of apoB RNA (33). All animals are allowed free access to water.

3.3. Isolation of Mouse Primary Hepatocytes

Upon anesthesia, a midline incision is made, the small intestine is gently deflected to the left side of the mouse, and kept in position with a saline soaked "all purpose sponge" exposing the portal vein. A single use sterile catheter connected to the perfusion solution prewarmed at 37°C is introduced into the vena cava and the portal vein is cut to allow the perfusate to flow out of the liver. During the entire procedure, the perfusion is performed at a flow rate of 3–4 mL/min and all perfusion solutions are maintained at 37°C. The liver is perfused with 35 mL perfusion working solution and massaged with Q-Tips dipped in saline to promote buffer flow until liver becomes pale (1–3 min). The liver is then perfused with Liver Digest Medium for 10 min followed by 3 min of perfusion working solution to wash out digest medium (see Note 2). Before removing the liver, the gallbladder is removed by cutting the fibrous band connecting the gallbladder to the diaphragm and the liver using a $4^{1/2}$" straight Iris scissors. The gallbladder is then disconnected from the bile duct by incision of the junction between the gallbladder and the bile duct, allowing the gallbladder to be removed intact. The digested liver is then gently teased apart with a cell scraper in 20 mL of L15 Medium containing 5% FBS and antibiotics. The homogenate is filtered through a 100 μm filter into a 50-mL centrifuge tube. Cells are collected by centrifugation at $50 \times g$ for 5 min at 4°C. The pelleted cells are washed twice with 40 mL hepatocyte wash medium and counted using a hematocytometer. Viability should be ~80%. Hepatocytes are then resuspended in culture medium and seeded at 6×10^5 cells/mL in a 35 mm collagen I coated dish. Hepatocytes are incubated at 37°C in a humidified atmosphere of 5% CO_2 for a period up to 5 days.

3.4. Protein and RNA Extraction

All liver tissues are minced and flash frozen in liquid nitrogen and stored at –80°C. Either whole small intestine or separate fractions (proximal, middle, and distal intestine) are minced and flash frozen similarly to the liver tissues. However, in some cases, it may be advantageous to scrape the mucosa in order to enrich for enterocytes rather than smooth muscle and supporting submucosal tissues. In that case, the content of the small intestine is flushed with 5–10 mL of ice-cold saline and the intestine is opened along the antimesenteric border to expose the mucosa that is then scraped on glass microscope slides. The scraped mucosa is then flash frozen in liquid nitrogen and stored at –80°C. Proteins are extracted from 100 mg tissue in 1 mL of lysis buffer. The homogenates are adjusted with 0.11 volume of 10× detergent buffer and incubated on ice for 10 min (see Note 3). Extracts are cleared from cellular debris by centrifugation at $16,000 \times g$ for 10 min. Protein concentrations in tissue homogenates are evaluated before being resolved on SDS-PAGE. RNAs are extracted by homogenization in TRIzol following the manufacturer's instructions and kept at –80°C (see Note 4).

3.5. Analysis of ACF and Apobec-1 mRNA Expression

Extracted RNAs (10 µg) are treated with two units of DNAse for 30 min at 37°C in 40 µL final volume to remove genomic DNA. One microgram of DNAse-treated RNA is added to 20 µL mixture containing Reverse Transcription buffer, 4 mM dNTP, 1× random primer and MultiScribe Reverse transcriptase, all provided in the High Capacity cDNA Reverse transcription kit. The reverse transcription reaction is performed as follows: step 1: 10 min at 25°C; step 2: 120 min at 37°C; step 3: 5 s at 85°C. The cDNAs are then kept at –80°C or used immediately for PCR amplification. Apobec-1 and ACF mRNA abundance is determined by quantitative PCR. Real-time quantitative PCRs are performed using SYBR Green Master mix following manufacturer's instructions. RNA abundance is determined by normalization against glyceraldehyde-3-phosphate dehydrogenase (GAPDH) level in each sample.

3.6. Evaluation of apoB C-to-U RNA Editing

3.6.1. In Vivo apoB RNA Editing by Poisoned Primer Extension

After RNA extraction and reverse transcription as described above, a 275-bp sequence of the murine apoB RNA (nt 6,512–6,786) flanking the editing site is amplified. PCR products are purified from the oligonucleotides using PCR purification kit and 20 ng of purified PCR product is annealed to 100 pg of 5′-end-labeled antisense primer located 39 nucleotides downstream of the edited site (see Note 5). The annealing reaction is performed at 70°C for 10 min. Primer extension is initiated by addition of 1.5 units of T7 RNA polymerase. The reaction is performed at 42°C for 3 min (3). The products are precipitated and resolved on an 8% polyacrylamide-7M urea gel (40 cm long) (see Note 6). The proportion of apoB48 and apoB100 is determined using Phosphor Imager/Image Quant).

3.6.2. Determination of apoB RNA Editing in S-100 Extracts

Tissues from liver, small intestine, and kidney are minced, homogenized sequentially in a Dounce homogenizer using a type A followed by a type B pestle, in Dignam A buffer, and centrifuged at $750 \times g$ in an SS34 rotor (Sorvall). The supernatants are adjusted with 0.11 volume of Dignam B buffer and centrifuged at $100,000 \times g$ in an SW55 Ti rotor (Beckman). After overnight dialysis at 4°C against 1,000–2,000 volumes of Dignam D buffer, the S100 extracts are stored at –80°C (3).

Exactly 10 fmol of *in vitro* transcribed 361 nt apoB RNA (nucleotides 6,512–6,872) is incubated with determined amounts of S100 extract (5–10 mg) for 3 h at 30°C in 1× *in vitro* conversion buffer supplemented with 10% glycerol, 250 ng tRNA, and 20 units RNAsin in a final volume of 20 μL. The reaction is stopped upon incubation with 200 ng/mL proteinase K. The RNA is extracted and subjected to RT-PCR followed by poisoned primer extension as described above. The proportion of edited and unedited apoB RNA is determined using PhosphorImager/ImageQuant.

3.7. Cellular Distribution of apobec-1 and ACF

3.7.1. Cellular Fractionation

Hepatic tissues (~100 mg) are minced in hypotonic buffer A, homogenized in Buffer B with a Dounce type B pestle.

The homogenate is centrifuged at $1,000 \times g$ for 5 min in an SS34 rotor. The supernatant is centrifuged further at $100,000 \times g$ yielding cytoplasmic extracts. The cytoplasmic extracts are dialyzed overnight at 4°C against Dignam D (1 or 2 L) and stored at –80°C. Pellets from the low-speed centrifugation are homogenized in Buffer C with a Dounce pestle B and centrifuged at $26,000 \times g$ for 10 min in an SS34 rotor resulting in nuclear extracts, kept at –80°C (34).

3.7.2. Western Blot Analysis of Apobec-1 and ACF Cellular Distribution

Cytoplasmic and nuclear extracts (~100 μg) are resolved on a 10% polyacrylamide-SDS gel and transferred to PVDF membranes. The blots are probed with rabbit polyclonal anti-apobec-1 antibody (35) or with rabbit polyclonal anti-ACF antibody (36). The blots are subsequently probed with cytoplasmic and nuclear-specific markers to confirm no cross contamination from one compartment to the other during subcellular fractionation (see Note 7).

3.7.3. Immunohistochemistry

Pieces (1–2 mm) of tissues (liver or small intestine) are fixed and paraffin-embedded. Immunohistochemistry is performed on 5-μm sections. The slides are rehydrated and then microwaved for 15 min in citrate buffer (0.01 M pH 6) before incubation with rabbit polyclonal ACF antibody (1:500 dilution) (36). For detection of apobec-1, slides are blocked for 15 min in 5% normal goat serum/3% bovine serum albumin, followed by exposure to rabbit anti-peptide apobec-1 antibody (40 mg/mL dilution) (35).

Following incubation with a biotinylated goat anti-rabbit IgG diluted in 0.3% Triton/1% BSA, the slides are treated for 15 min with 1% hydrogen peroxide in methanol to block endogenous peroxidases. Finally, the slides are treated with avidin-conjugated horseradish peroxidase prior to color development.

3.8. Serum apoB Analysis

Blood samples are drawn by exsanguination via direct cardiac puncture or aortic cannulation or by retro-orbital bleed, and serum is obtained by centrifugation at $4,000 \times g$ for 20 min at 4°C. Two microliters of serum is resolved on a 4–12% polyacrylamide gradient-SDS gel. After transfer, membranes are incubated with rabbit polyclonal anti-mouse apoB antibody (37) and visualized by enhanced chemoluminescence.

3.9. Analysis of Serum Lipids and Lipoproteins Content

Serum triglycerides are analyzed using L-type TG H kit (Wako Diagnostics, L type TG-M-ColorA, L type TG-M-ColorB). Serum cholesterol is assessed using Cholesterol E kit. Lipoproteins in 200 μL of plasma are fractionated by Superose 6 chromatography. Specifically, two 25-cm Superose 6 columns are connected in tandem and 500 μL fractions are collected. Samples are either stored at 4°C for up to 2 weeks or kept at –20°C until analysis. Cholesterol and triglycerides content of each individual fraction can be assessed using enzymatic detection kits described above.

4. Notes

1. Two-to-four-month-old, tetracycline-dependent, liver-specific conditional apobec-1 transgenic mice are administered 10 mg/mL doxycycline hydrochloride in the drinking water. The expression of rabbit apobec-1 is suppressed within 2 days and followed by suppression of apoB RNA editing activity and apoB48 synthesis within 4 days of treatment.

2. Time of digestion is critical for the quality of the hepatocyte preparation Consequently, it is recommended to adjust the duration of digestion to the size of the liver. Small livers will be sufficiently digested in 6–7 min. Check the texture of the liver: the surface should evolve from a smooth and organized liver to a liver with a dissociated appearance in which cellular components can be distinguished.

3. The detergent is added to the crude protein extract after homogenization to avoid too much foaming. Add detergent and vortex before the incubation on ice.

4. Use 1 mL of TRIzol per 100 mg of tissue.

5. 5′-end-labeled primer is purified from the unincorporated nucleotide using QIAquick Nucleotide Removal kit (QIAGEN).

6. Resuspend the poisoned primer extension products in 3 μL of loading dye containing 95% formamide, 0.025% xylene cyanol, 0.025% bromophenol blue, 18 mM EDTA, 0.025% SDSs and incubate at 95°C for 4 min. Run the gel at 70 W and stop the electrophoresis when the Bromophenol blue reaches the two-thirds of the gel.

7. Primary antibody solutions can be kept at 4°C for future experiments.

Acknowledgments

Work cited in this review was supported by grants from the National Institutes of Health (HL-38180, DK-56260, DK-52574) to NOD. The authors are deeply grateful to Susan Kennedy and Jianyang Luo for assistance with the murine models quoted in this review.

References

1. Blanc, V., and Davidson, N. O. (2003) C-to-U RNA editing: mechanisms leading to genetic diversity. *J Biol Chem* **278**, 1395–8.

2. Greeve, J., Altkemper, I., Dieterich, J. H., Greten, H., and Windler, E. (1993) Apolipoprotein B mRNA editing in 12 different mammalian species: hepatic expression is reflected in low concentrations of apoB-containing plasma lipoproteins. *J Lipid Res* **34**, 1367–83.

3. Giannoni, F., Bonen, D. K., Funahashi, T., Hadjiagapiou, C., Burant, C. F., and Davidson, N. O. (1994) Complementation of apolipoprotein B mRNA editing by human liver accompanied by secretion of apolipoprotein B48. *J Biol Chem* **269**, 5932–6.

4. Davidson, N. O., and Shelness, G. S. (2000) Apolipoprotein B:mRNA editing, lipoprotein assembly and presecretory degradation. *Annu Rev Nutr* **20**, 169–93.

5. Teng, B., Burant, C. F., and Davidson, N. O. (1993) Molecular cloning of an apolipoprotein B messenger RNA editing protein. *Science* **260**, 1816–9.

6. Mehta, A., Kinter, M. T., Sherman, N. E., and Driscoll, D. M. (2000) Molecular cloning of apobec-1 complementation factor, a novel RNA-binding protein involved in the editing of apolipoprotein B mRNA. *Mol Cell Biol* **20**, 1846–54.

7. Lellek, H., Kirsten, R., Diehl, I., Apostel, F., Buck, F., and Greeve, J. (2000) Purification and molecular cloning of a novel essential component of the apolipoprotein B mRNA editing enzyme-complex. *J Biol Chem* **275**, 19848–56.

8. Blanc, V., Henderson, J. O., Newberry, E. P., Kennedy, S., Luo, J., and Davidson, N. O. (2005) Targeted deletion of the murine apobec-1 complementation factor (acf) gene results in embryonic lethality. *Mol Cell Biol* **25**, 7260–9.

9. Teng, B., Black, D. D., and Davidson, N. O. (1990) Apolipoprotein B messenger RNA editing is developmentally regulated in pig small intestine: nucleotide comparison of apolipoprotein B editing regions in five species. *Biochem Biophys Res Commun* **173**, 74–80.

10. Henderson, J. O., Blanc, V., and Davidson, N. O. (2001) Isolation, characterization and developmental regulation of the human apobec-1 complementation factor (ACF) gene. *Biochim Biophys Acta* **1522**, 22–30.

11. Mansouri, A., Chowdhury, K., and Gruss, P. (1998) Follicular cells of the thyroid gland require *Pax8* gene function. *Nat Genet* **19**, 87–90.

12. Flamant, F., Poguet, A. L., Plateroti, M., Chassande, O., Gauthier, K., Streichenberger, N., Mansouri, A., and Samarut, J. (2002) Congenital hypothyroid Pax8(−/−) mutant mice can be rescued by inactivating the TRalpha gene. *Mol Endocrinol* **16**, 24–32.

13. Mukhopadhyay, D., Plateroti, M., Anant, S., Nassir, F., Samarut, J., and Davidson, N. O. (2003) Thyroid hormone regulates hepatic triglyceride mobilization and apolipoprotein B messenger ribonucleic acid editing in a murine model of congenital hypothyroidism. *Endocrinology* **144**, 711–9.

14. Philips, M. S., Liu, Q., Hammond, H. A., Dugan, V., Hey, P. J., Caskey, C. T., and Hess, J. F. (1996) Leptin receptor missense in the fatty Zucker rat. *Nature Genet* **13**, 18–19.

15. von Wronski, M. A., Hirano, K. I., Cagen, L. M., Wilcox, H. G., Raghow, R., Thorngate, F. E., Heimberg, M., Davidson, N. O., and Elam, M. B. (1998) Insulin increases expression of apobec-1, the catalytic subunit of the apolipoprotein B mRNA editing complex in rat hepatocytes. *Metabolism* **47**, 869–73.

16. Elam, M. B., von Wronski, M. A., Cagen, L., Thorngate, F., Kumar, P., Heimberg, M., and Wilcox, H. G. (1999) Apolipoprotein B mRNA editing and apolipoprotein gene expression in the liver of hyperinsulinemic fatty Zucker rats: relationship to very low density lipoprotein composition. *Lipids* **34**, 809–16.

17. Sowden, M. P., Ballatori, N., Jensen, K. L., Reed, L. H., and Smith, H. C. (2002) The editosome for cytidine to uridine mRNA editing has a native complexity of 27S: identification of intracellular domains containing active and inactive editing factors. *J Cell Sci* **115**, 1027–39.

18. Lehmann, D. M., Galloway, C. A., Sowden, M. P., and Smith, H. C. (2006) Metabolic regulation of apoB mRNA editing is associated with phosphorylation of APOBEC-1 complementation factor. *Nucleic Acids Res* **34**, 3299–308.

19. Wiegman, C. H., Bandsma, R. H., Ouwens, M., van der Sluijs, F. H., Havinga, R., Boer, T., Reijngoud, D. J., Romijn, J. A., and Kuipers, F. (2003) Hepatic VLDL production in *ob/ob* mice is not stimulated by massive *de novo* lipogenesis but is less sensitive to the suppressive effects of insulin. *Diabetes* **52**, 1081–9.

20. Lee, G. H., Proenca, R., Montez, J. M., Carroll, K. M., Darvishzadeh, J. G., Lee, J. I., and Friedman, J. M. (1996) Abnormal splicing of the leptin receptor in diabetic mice. *Nature* **379**, 632–5.

21. Leiter, E. H., Coleman, D. L., Eisenstein, A. B., and Strack, I. (1980) A new mutation (db3j) at the diabetes locus in strain 129/j mice. Physiological and histological characterization. *Diabetologia* **19**, 58–65.

22. Kobayashi, K., Forte, T. M., Taniguchi, S., Ishida, B. Y., Oka, K., and Chan, L. (2000) The *db/db* mouse, a model for diabetic dyslipidemia: molecular characterization and effects of Western diet feeding. *Metabolism* **49**, 22–31.

23. Hirano, K., Young, S. G., Farese, R. V., Jr., Ng, J., Sande, E., Warburton, C., Powell-Braxton, L. M., and Davidson, N. O. (1996) Targeted disruption of the mouse apobec-1 gene abolishes apolipoprotein B mRNA editing and eliminates apolipoprotein B48. *J Biol Chem* **271**, 9887–90.

24. Nakamuta, M., Chang, B. H., Zsigmond, E., Kobayashi, K., Lei, H., Ishida, B. Y., Oka, K., Li, E., and Chan, L. (1996) Complete phenotypic characterization of apobec-1 knockout mice with a wild-type genetic background and a human apolipoprotein B trasngenic background and restoration of apolipoprotein B mRNA editing by somatic gene transfer of Apobec-1. *J Biol Chem* **271**, 25981–8.

25. Yamanaka, S., Balestra, M. E., Ferrell, L. D., Fan, J., Arnold, K. S., Taylor, S., Taylor, J. M., and Innerarity, T. L. (1995) Apolipoprotein B mRNA-editing protein induces hepatocellular carcinoma and dysplasia in transgenic animals. *Proc Natl Acad Sci U S A* **92**, 8483–7.

26. Yamanaka, S., Poksay, K. S., Driscoll, D. M., and Innerarity, T. L. (1996) Hyperediting of multiple cytidines of apolipoprotein B mRNA by APOBEC-1 requires auxiliary protein(s) but not a mooring sequence motif. *J Biol Chem* **271**, 11506–10.

27. Kistner, A., Gossen, M., Zimmermann, F., Jerecic, J., Ullmer, C., Lubbert, H., and Bujard, H. (1996) Doxycycline-meiated quantitative and tissue-specific control of gene expression in transgenic mice. *Proc Natl Acad Sci U S A* **93**, 10933–8.

28. Hersberger, M., Patarroyo-White, S., Qian, X., Arnold, K. S., Rohrer, L., Balestra, M. E., and Innerarity, T. L. (2003) Regulatable liver expression of the rabbit apolipoprotein B mRNA-editing catalytic polypeptide 1 (APOBEC-1) in mice lacking endogenous APOBEC-1 leads to aberrant hyperediting. *Biochem J* **369**, 255–62.

29. Davidson, N. O., Powell, L. M., Wallis, S. C., and Scott, J. (1988) Thyroid hormone modulates the introduction of a stop codon in rat liver apolipoprotein B messenger RNA. *J Biol Chem* **263**, 13482–5.

30. Plateroti, M., Chassande, O., Fraichard, A., Gauthier, K., Freund, J. N., Samarut, J., and Kedinger, M. (1999) Involvement of T3Ralpha-and beta-receptor subtypes in mediation of T3 functions during postnatal murine intestinal development. *Gastroenterology* **116**, 1367–78.

31. Srivastava, R. A. (1995) Increased apoB100 mRNA in inbred strains of mice by estrogen is caused by decreased RNA editing protein mRNA. *Biochem Biophys Res Commun* **212**, 381–7.

32. Lau, P. P., Cahill, D. J., Zhu, H. J., and Chan, L. (1995) Ethanol modulates apolipoprotein B mRNA editing in the rat. *J Lipid Res* **36**, 2069–78.

33. Baum, C. L., Teng, B. B., and Davidson, N. O. (1990) Apolipoprotein B messenger RNA editing in the rat liver. Modulation by fasting and refeeding a high carbohydrate diet. *J Biol Chem* **265**, 19263–70.

34. MacGinnitie, A. J., Anant, S., and Davidson, N. O. (1995) Mutagenesis of apobec-1, the catalytic subunit of the mammalian apolipoprotein B mRNA editing enzyme reveals distinct domains that mediate cytosine nucleoside deaminase, RNA binding and RNA editing activity. *J Biol Chem* **270**, 14768–75.

35. Funahashi, T., Giannoni, F., DePaoli, A. M., Skarosi, S. F., and Davidson, N. O. (1995) Tissue-specific, developmental and nutritional regulation of the gene encoding the catalytic subunit of the rat apolipoprotein B mRNA editing enzyme: functional role in the modulation of apoB mRNA editing. *J Lipid Res* **36**, 414–28.

36. Blanc, V., Henderson, J. O., Kennedy, S., and Davidson, N. O. (2001) Mutagenesis of apobec-1 complementation factor reveals distinct domains that modulate RNA binding, protein-protein interaction with apobec-1 and complementation of C to U RNA editing activity. *J Biol Chem* **276**, 46386–93.

37. Bonen, D. K., Nassir, F., Hausman, A. M., and Davidson, N. O. (1998) Inhibition of N-linked glycosylation results in retention of intracellular apo(a) in hepatoma cells, although nonglycosylated and immature forms of apolipoprotein(a) are competent to associate with apolipoprotein B-100 in vitro. *J Lipid Res* **39**, 1629–40.

Chapter 8

In Vivo Analysis of RNA Editing in Plastids

Stephanie Ruf and Ralph Bock

Abstract

mRNA editing in plastids (chloroplasts) of higher plants proceeds by cytidine-to-uridine conversion at highly specific sites. Editing sites are recognized by the interplay of *cis*-acting elements at the RNA level and site-specific *trans*-acting protein factors that are believed to bind to the *cis*-elements in a sequence-specific manner. The C-to-U editing enzyme, a presumptive cytidine deaminase acting on polynucleotides, is still unknown. The development of methods for the stable genetic transformation of the plastid genome in higher plants has facilitated the analysis of RNA editing *in vivo*. Plastid transformation has been extensively used to define the sequence requirements for editing site selection and to address questions about editing site evolution. This chapter describes the basic methods involved in the generation and analysis of plants with transgenic chloroplast genomes and summarizes the applications of plastid transformation in editing research.

Key words: RNA editing, Plastid, Chloroplast, *Nicotiana tabacum*, Plastid transformation, Biolistic transformation, Particle gun, *cis*-Acting element, Evolution

1. Introduction

Plastids (chloroplasts) of higher plants exhibit two types of conversional RNA editing: cytidine-to-uridine editing in mRNAs (1–3) and adenosine-to-inosine editing in a plastid genome-encoded tRNA, the tRNA-Arg(ACG) (4). C-to-U editing occurs also in the other DNA-containing organelle of the plant cell, the mitochondrion (2). Plastid mRNA editing is a posttranscriptional process and largely independent of chloroplast translation, intron splicing, and cleavage of polycistronic precursors into mature monocistronic mRNAs (5–8). With very few exceptions (9), plastid C-to-U editing sites reside within coding regions and are functionally significant in that the C-to-U change alters the coding properties of the affected triplets (10). In most instances, these

Ruslan Aphasizhev (ed.), *RNA and DNA Editing: Methods and Protocols*, Methods in Molecular Biology, vol. 718,
DOI 10.1007/978-1-61779-018-8_8, © Springer Science+Business Media, LLC 2011

alterations result in the restoration of codons for phylogenetically conserved amino acid residues (10–13). At most sites, mRNA editing is virtually complete in that unedited precursors are hardly detectable. At some sites, the editing is subject to developmental and/or tissue-specific variation in efficiency (14–16), although the functional relevance of this regulation is still unclear. Plastid (and also mitochondrial) C-to-U editing has appeared relatively late in plant evolution. It is restricted to vascular plants (tracheophytes) and some derived bryophyte lineages, but does not occur in algae and cyanobacteria, the evolutionary ancestors of all present-day plastids (2, 12). Interestingly, many plastid RNA editing sites are only poorly conserved between species (10, 12, 13) indicating that editing sites are frequently lost (by acquisition of a genomic C-to-T mutation) and/or have been gained several times independently during plant evolution.

RNA processing by C-to-U editing can be essential for normal protein function, as demonstrated by the mutant phenotype of a genetically engineered tobacco plant harboring a non-editable version of a chloroplast genome-encoded photosynthesis gene (17). The molecular recognition of plastid mRNA editing sites is mediated by a *cis*-acting RNA element residing in the region immediately upstream of the editing site (18–22) and *trans*-acting protein factor(s) binding to the *cis*-element (23–25). Recently, members of the pentatricopeptide repeat (PPR) protein family have emerged as *trans*-factors involved in the specific recognition of mRNA editing sites in plastids (26–28). However, while the chloroplast adenosine-to-inosine tRNA editing enzyme was recently identified (4), the identity of the C-to-U mRNA editing enzyme – a putative cytidine deaminase capable of acting on polynucleotide substrates – has not yet been revealed.

Research into the molecular mechanisms of plastid mRNA editing has mainly relied on two methodologies: (1) *in vivo* analyses by generating plants with transgenic plastids (17, 23, 29) and (2) *in vitro* assays using RNA editing-competent chloroplast extracts (22, 25, 30–33). This chapter focuses on the methods involved in the *in vivo* analysis of RNA editing by transformation of the plastid genome (34, 35) in the model plant tobacco, *Nicotiana tabacum*. The generation of plants with transgenic chloroplasts, also referred to as transplastomic plants (36–39), has been extensively used in plant editing research to (1) identify *cis*-acting elements at the RNA level which are involved in editing site selection (18–22, 40) and (2) transplant RNA editing sites from one species into another (17, 41–44) in order to test for evolutionary conservation of site-specific *trans*-acting editing factors (Fig. 1). Here we describe the procedures involved in the design of transformation vectors for editing research, the selection of transplastomic tobacco plants, and their molecular characterization and provide updated protocols (35) for plastid transformation by particle gun-mediated transformation.

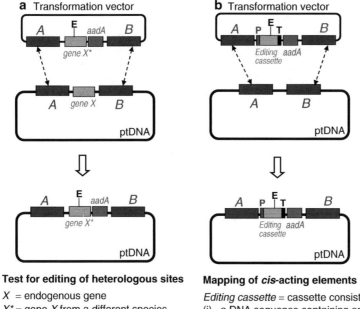

a Transformation vector

b Transformation vector

Test for editing of heterologous sites

X = endogenous gene
X^* = gene X from a different species
    (containing an RNA editing site, E)

Mapping of *cis*-acting elements

Editing cassette = cassette consisting of
(i) a DNA sequence containing an RNA
     editing site (E),
(ii) a plastid promoter (P),
(iii) a 3' UTR sequence (terminator, T)

Fig. 1. Applications of the transplastomic technology in RNA editing research and design of plastid transformation vectors. (**a**) Introduction of a heterologous editing site into an endogenous plastid gene by gene replacement. The editing site-containing allele is linked to the *aadA* selectable marker gene to replace the endogenous allele by homologous recombination. (**b**) Plastid transformation vector suitable for mapping of *cis*-acting sequence elements for RNA editing site recognition. For expression of an editing site-containing fragment in a chimeric sequence context, the editing cassette is physically linked to the *aadA* gene and targeted to a neutral insertion site, typically an intergenic region in the plastid genome. *A,B* flanking plastid DNA regions for integration via homologous recombination; *ptDNA* plastid genome; *UTR* untranslated region.

2. Materials

All chemicals used in synthetic media for plant tissue culture must be of highest purity, and ideally, should be certified for plant cell and tissue culture.

2.1. Chemicals and Stock Solutions for Plant Culture Media

1. 10× Macro (1 L): 19.0 g KNO_3, 16.5 g NH_4NO_3, 4.4 g $CaCl_2 \cdot 2H_2O$, 3.7 g $MgSO_4 \cdot 7H_2O$, 1.7 g KH_2PO_4.

2. 100× Micro (100 ml): 169 mg $MnSO_4 \cdot 2H_2O$, 86 mg $ZnSO_4 \cdot 7H_2O$, 62 mg H_3BO_3, 8.30 mg KI, 2.50 mg $Na_2MoO_4 \cdot 2H_2O$, 0.25 mg $CuSO_4 \cdot 5H_2O$, 0.25 mg $CoCl_2 \cdot 6H_2O$.

3. FeNaEDTA: 1 g/100 ml.

4. Thiamine HCl: 1 mg/ml.

5. 1-Naphthylacetic acid (NAA): 1 mg/ml (in 0.1 M NaOH).

6. N^6-benzylamino purine (BAP): 1 mg/ml (in 0.1 M HCl).

Sterilization of 10× Macro and 100× Micro stock solutions is performed by autoclaving, all other stock solutions by sterile filtration. As an alternative to the preparation of these nutrient and vitamin stock solutions, various salt mixtures and ready-to-use media are commercially available (e.g., from Duchefa, Haarlem, The Netherlands).

2.2. Plant Culture Media

1. MS medium (45) for aseptic growth of tobacco plants (1 L): 10× Macro (100 ml), 100× Micro (10 ml), FeNaEDTA (5 ml), 30 g sucrose, pH 5.6–5.8 (adjust with 0.1 M KOH).

   The liquid medium is supplemented with 7.4 g/L agar (certified for plant cell and tissue culture; e.g., Micro Agar, Duchefa, Haarlem, The Netherlands; see Note 1), autoclaved and poured in Petri dishes or sterile containers for plant growth.

2. RMOP medium for plant regeneration from tissue explants.

   Regeneration of shoots from tissue explants is induced by addition of the synthetic plant hormones NAA (a structural analog of auxin) and BAP (a cytokinin analog) to the basic MS medium.

   RMOP medium (1 L): 10× Macro (100 ml), 100× Micro (10 ml), FeNaEDTA (5 ml), 30 g sucrose, thiamine HCl (1 ml), NAA (0.1 ml), BAP (1 ml), 100 mg myo-inositol, pH 5.8 (0.2 M KOH). After addition of 7.4 g/L agar (for plant cell and tissue culture; Duchefa; see Note 1), the medium is sterilized by autoclaving and chilled to approximately 60°C. Subsequently, antibiotics are added if required and the medium is poured into Petri dishes (approximately 50 ml medium per Petri dish of 9 cm diameter and 2 cm height).

2.3. Particle Preparation

1. Gold particles (0.6 μm, Bio-Rad).
2. Ethanol 100%.
3. Spermidine (free base).
4. $CaCl_2$.

3. Methods

Currently, plastid transformation is routine only in relatively few higher plant species, including tobacco (36, 46), tomato (47, 48), and soybean (49, 50). Tobacco represents the by far most efficient and least time-consuming experimental system for plastid transformation. Therefore, it is highly recommended to at least initially use tobacco to set up the technology and acquire the basic expertise. Particle gun-mediated transformation, also referred to as "biolistic" transformation (Fig. 2, ref. (51)), represents the

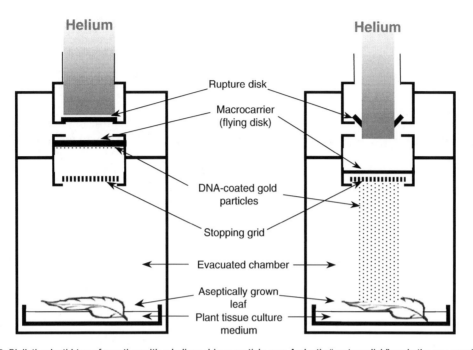

Fig. 2. Biolistic plastid transformation with a helium-driven particle gun. A plastic "rupture disk" seals the gas accelera-
tion tube (*left*). The gold particles coated with the transforming plasmid DNA are spread on the surface of a second plastic
disk ("macrocarrier"). A metal grid ("stopping grid") determines the flight distance of the macrocarrier. When the helium
pressure reaches the burst pressure of the rupture disk, the disk breaks (*right*) and the gas accelerates the macrocarrier
with the particles on its bottom side. When the flying disk hits the metal grid, the DNA-coated particles fly through the
meshes of the grid and penetrate the cells of the sterile tobacco leaf placed in a Petri dish at the bottom of the evacuated
chamber (modified from ref. (*35*)).

currently most effective method of delivering foreign DNA into
higher plant plastids (34, 46, 52). Polyethylene glycol (PEG)
treatment of protoplasts in the presence of transforming DNA
provides an alternative (53, 54), but the cell culture and transformation procedures involved in PEG-mediated plastid transformation are considerably more demanding and time-consuming than
biolistic transformation.

3.1. Plant Material and Growth Conditions

Young leaves from tobacco plants (*N. tabacum*) grown under
aseptic conditions are used for chloroplast transformation
experiments. Such sterile tobacco plants can be readily obtained
from surface-sterilized seeds. To this end, 400 μl ethanol and
400 μl 6% bleach (sodium hypochlorite solution) are added to
a sample of 50–100 seeds in a 1.5-ml Eppendorf tube. Following
vigorous shaking for 2–3 min, the liquid is removed with a sterile
pipet and the seeds are washed at least four times with 1.5-ml
sterile distilled water (each time removing the liquid as completely as possible). Seeds are then transferred to sterile containers (e.g., Magenta boxes) and germinated on MS medium

(see below). Individual plants should then be pricked out and raised in boxes on MS medium in a growth chamber at 25°C under a 16-h light/8-h dark cycle until they reach a height of approximately 10 cm. The light intensity should be between 50 and 100 μE/m²s.

3.2. Design of Plastid Transformation Vectors

The tobacco plastid transformation technology has been extensively used to map *cis*-acting elements involved in editing site recognition (21, 23), to define minimum RNA substrates for editing *in vivo* (18, 19), and to test for editing of heterologous sites (i.e., sites taken from plastid genomes of other species or from plant mitochondrial genomes; refs. (17, 41, 44)). The principles of vector design for these applications are illustrated in Fig. 1.

To test heterologous sites from other species for their editability in tobacco plastids, gene replacement represents the most appropriate strategy (Fig. 1a). The efficient homologous recombination system present in chloroplasts facilitates gene-targeting approaches and makes the creation of an RNA editing site by site-directed mutagenesis (T-to-C change) or by exchange of a restriction fragment rather straightforward (17, 44). The mutant allele is linked to the selectable marker gene *aadA* (a spectinomycin/streptomycin resistance gene encoding an antibiotic-inactivating enzyme; ref. (46)) and two homologous recombination events in the flanking regions (*A* and *B* in Fig. 1) exchange the endogenous plastid gene by the foreign DNA in vivo. It is important to realize that, in such an experiment, the linkage between the selectable antibiotic resistance marker (*aadA* gene; Fig. 1) and the editing site to be introduced is incomplete. Thus, if homologous recombination is resolved between the *aadA* gene and the editing site, only the selectable maker gene will be incorporated into the plastid genome, but not the editing site. The probability of this happening depends on the physical distance between the *aadA* and the editing site (17, 29, 36), but if a sufficient number of plastid transformants is obtained (5–10), even transplastomic lines with distant point mutations can be recovered at reasonable frequencies (17, 29).

To identify *cis*-acting recognition elements or to test mutated editing substrates, editing site-containing DNA fragments are placed into a chimeric context (an "editing cassette"; Fig. 1b) and expressed as minigenes (18, 19, 21, 55). Transformation of such a minigene construct into chloroplasts introduces an additional copy of the respective editing site into the plastid genome. Strong overexpression of an editing site-containing RNA can lead to competition with the resident editing site(s) for *trans*-factors (23). This can provide valuable additional information by identifying groups of editing sites that share common *trans*-acting factors (55, 56). Expression of the minigene can be adjusted to the desired level by choosing an expression cassette of suitable strength (e.g., ref. (57–62)). Serial deletions and/or scanning point mutagenesis of editing site-containing fragments can be used to systematically dissect the sequence requirements for editing site recognition (18–21).

3.3. Gold Particle Preparation for Biolistic Plastid Transformation

All steps should be performed on ice or in the cold room (at 4°C). Water and 100% ethanol must be ice-cold.

1. Use 2 mg of gold particles (0.6 μm diameter) per 10 shots.
2. Prepare a gold suspension: $n \times 2$ mg in $n \times 100$ μl of 100% EtOH.
3. Vortex at least 1 min at maximum power.
4. Centrifuge 1 s at 4,000 rpm in a microfuge.
5. Remove the supernatant completely and redissolve the particles in 1 ml of sterile, distilled water.
6. Centrifuge 1 s at 4,000 rpm in a microfuge, remove, and discard the supernatant.
7. Resuspend the particles very carefully in $n \times 250$ μl sterile water.
8. Make 245 μl aliquots (per 10 shots).
9. Add to each aliquot in the following order: 20 μg of DNA (plastid transformation vector; concentration 1–2 μg/μl) 250 μl of 2.5 M $CaCl_2$ 50 μl of 100 mM spermidine. (Vortex briefly immediately after addition of each component.)
10. Incubate on ice for 10 min with brief vortexing every minute.
11. Centrifuge 1 s at 3,500 rpm in a microfuge and remove the supernatant completely.
12. Add 600 μl of 100% EtOH and carefully resuspend the particles by pipetting and vigorous vortexing.
13. Centrifuge 1 s at 3,500 rpm in a microfuge and remove the supernatant completely.
14. Repeat this washing step.
15. Resuspend the particles very carefully in 65 μl 100% EtOH by pipetting and vigorous vortexing (see Note 2).
16. Use 6 μl/shot and carefully resuspend particles immediately before use.

3.4. Biolistic Bombardment

Use young leaves from tobacco plants (*N. tabacum*) grown under aseptic conditions on synthetic medium (MS medium). Place the leaf in a Petri dish onto a sterile piece of filter paper on top of a thin layer of RMOP medium without antibiotics. Bombard the abaxial side of the leaf (see Note 3) with the plasmid DNA-coated gold particles. For the Bio-Rad PDS1000He biolistic gun (Fig. 2), suitable settings for the transformation of tobacco chloroplasts are as follows:

- Helium pressure at the tank regulator: 1,300–1,400 psi
- Rupture disks: 1,100 psi (Bio-Rad)
- Macrocarrier (flying disk) assembly: level two from the top
- Petri dish holder: level two from the bottom
- Vacuum (at the time of the shot): 26–28 in. Hg

A typical chloroplast transformation experiment involves the bombardment of 10–15 tobacco leaves per DNA construct. Alternatively, the Hepta Adaptor setup of the Bio-Rad PDS1000He particle gun can be used to bombard a standard Petri dish (9 cm diameter) that is fully covered with leaf material (48). This reduces the required number of shots by a factor of approximately 4–5.

To adjust the settings of the particle gun, optimize the particle preparation procedure, and control for its quality, transient nuclear transformation assays using a GUS (β-glucuronidase) reporter gene construct should be used. Any plant transformation vector containing a GUS reporter gene under the control of a strong constitutive promoter can be used for these assays (see Note 4). Assessment of transient gene expression by histochemical GUS staining (35) provides a suitable proxy also for stable plastid transformation efficiency.

3.5. Selection of Transplastomic Plants

Following biolistic bombardment, the leaves are cut into small pieces (approximately 5×5 mm) and placed adaxial side up onto the surface of a selective regeneration medium (RMOP containing 500 μg/ml spectinomycin). Resistant calli or shoots typically appear after 3–6 weeks of incubation (light intensity: approximately 50 μE/m²s; 16-h light/8-h dark cycle). Note that not all of the antibiotic-resistant plant lines are true plastid transformants. Spontaneous spectinomycin resistance can be conferred by specific point mutations in the chloroplast 16S rRNA gene, which arise spontaneously at low frequency (63). However, these point mutations act in a strictly antibiotic-specific manner; spontaneous spectinomycin-resistant mutants are sensitive to streptomycin and vice versa. In contrast, the *aadA* selectable marker gene confers broad resistance to a wider range of aminoglycoside-type antibiotics, including spectinomycin and streptomycin. Consequently, when exposed to double selection on plant regeneration medium (RMOP medium) containing both spectinomycin and streptomycin, tissue samples from spontaneous spectinomycin-resistant lines bleach out, whereas true plastid transformants remain green, display continued callus growth, and ultimately regenerate into shoots (36). Alternatively, true chloroplast transformants can be identified by DNA gel blot analysis or PCR assays, which test for the physical presence of the resistance gene in the plastid genome (17, 18).

Plant cells are highly polyploid with respect to their plastid genome, and a single tobacco leaf cell contains thousands of identical copies of the plastid DNA distributed among at least dozens of chloroplasts. For this reason, primary plastid transformants are usually heteroplasmic, meaning that they contain a mix of wild-type and transformed genome copies. In order to obtain a genetically stable transplastomic plant, residual wild-type genome molecules must be completely eliminated. This is achieved by subjecting the primary plastid transformant to additional cycles of regeneration on selective medium (36). Typically after two to three rounds, homoplasmic transplastomic lines are obtained, in which the wild-type genome is no longer detectable. Homoplasmy is assessed by either DNA gel blot analysis (46) or sensitive PCR assays that selectively favor amplification of residual wild-type genomes (18). Verification of homoplasmy is particularly important when the transplastomic lines can be expected to display a mutant phenotype (17) or when the lines are to be used for quantitative analyses of editing efficiency.

It is important to note that homoplasmic transplastomic lines sometimes still show faint wild-type-like hybridization signals in Southern blots (for two recent examples, see refs. (64, 65)). These bands are not necessarily indicative of the presence of residual wild-type copies of the plastid genome. Instead, they often represent the so-called promiscuous DNA: non-functional copies of the plastid DNA residing in the nuclear or mitochondrial genome (66–68). To distinguish between residual heteroplasmy and promiscuous DNA, either Southern blots with purified plastid DNA (69, 70) or inheritance assays (seed tests; see below) should be conducted.

Finally, homoplasmic, transplastomic shoots regenerated on RMOP medium are transferred to boxes with MS medium. The phytohormone-free MS medium will induce rooting, and the plants can then be taken out of the sterile environment, transferred to soil (see Note 5), and grown to maturity in the greenhouse (see Note 6).

3.6. Analysis of Transplastomic Plants

As chloroplasts are maternally inherited in tobacco (like in the vast majority of angiosperm plant species; refs. (71, 72)), plastid transgene localization and homoplasmy can be easily verified genetically by conducting reciprocal crosses. To this end, wild-type and transformed plants are transferred to soil and grown to maturity in the greenhouse. Seed pods are collected from selfed plants and from reciprocal crosses of the transplastomic lines with the wild type. Surface-sterilized seeds are then germinated on spectinomycin- or streptomycin-containing (500 mg/L) MS medium and analyzed for uniparental inheritance of the plastid-encoded antibiotic resistance. Selfed transformants and crosses with the plastid transformant as maternal parent give rise to uniformly green (i.e., drug-resistant) progeny, whereas seeds collected from wild-type plants and crosses with the plastid transformant as paternal parent yield white (i.e., antibiotic-sensitive) seedlings (36). If white seedlings turn up in the inheritance assay with seeds from the selfed transplastomic plant, the plant was still heteroplasmic and should be purified further until homoplasmy is attained.

All methods involved in analyzing RNA editing in transplastomic plants represent standard molecular biology procedures. Simple protocols are available that allow the simultaneous isolation of high-quality DNA and RNA suitable for reverse transcription (e.g., ref. (73)). RNA editing efficiency can be quantitated using conventional methods (18, 34, 40, 74).

4. Notes

1. The exact amount of agar is dependent on the autoclave used (and, in particular, on its heating and cooling speeds), and therefore must be determined empirically. The solidified medium should be soft enough to allow pushing of leaf pieces 1–2 mm into the medium (for optimum medium contact to induce regeneration), but hard enough to stably hold an inserted stem cutting in a vertical position.

2. DNA-coated gold particles tend to aggregate. Careful resuspension of the particles is crucial to the success of the transformation experiment. Clumpy particles hit fewer target cells and will cause severe cell damage upon penetrating the leaf, thus greatly reducing both transformation efficiency and plant regeneration frequency.

3. The abaxial (lower) side of the leaf usually has a softer epidermis and cuticle than the adaxial (upper) side and thus is more efficiently penetrated by the particles.

4. Vectors for stable *Agrobacterium*-mediated plant transformation (e.g., pBI121) are often relatively large plasmids and, therefore, may give somewhat lower transient transformation frequencies than smaller vectors for biolistic transformation or protoplast transformation. A reasonably small GUS gene-containing vector suitable for transient transformation assays is, for example, vector pFF19G (75).

5. Residual culture medium should be washed off from the roots of the plants as completely as possible. As the medium is rich in minerals and also contains sucrose, it can be rapidly colonized by soil microbes, including plant root-infecting pathogens.

6. Plants raised in sterile culture containers are adapted to 100% relative humidity and, therefore, are extremely sensitive to dry air. To allow for gradual adaptation to greenhouse conditions, plants should be kept under a transparent plastic cover for at least a few days which is then stepwise lifted.

Acknowledgments

Work on RNA editing and plastid transformation in the authors' laboratory is supported by the Max Planck Society and by grants from the Deutsche Forschungsgemeinschaft (DFG), the Bundesministerium für Bildung und Forschung (BMBF), and the European Union (Framework Programs 6 and 7).

References

1. Bock, R. (2000) Sense from nonsense: how the genetic information of chloroplasts is altered by RNA editing. *Biochimie* **82**, 549–557.

2. Bock, R. (2001) RNA editing in plant mitochondria and chloroplasts. *Frontiers in Molecular Biology: RNA Editing, Bass, B. (ed.), Oxford University Press, New York,* 38–60.

3. Schmitz-Linneweber, C. and Barkan, A. (2007) RNA splicing and RNA editing in chloroplasts. *Top. Curr. Genet.* **19**, 213–248.

4. Karcher, D. and Bock, R. (2009) Identification of the chloroplast adenosine-to-inosine tRNA editing enzyme. *RNA* **15**, 1251–1257.

5. Ruf, S., Zeltz, P. and Kössel, H. (1994) Complete RNA editing of unspliced and dicistronic transcripts of the intron-containing reading frame IRF170 from maize chloroplasts. *Proc. Natl. Acad. Sci. U S A* **91**, 2295–2299.

6. Zeltz, P., Hess, W. R., Neckermann, K., Börner, T. and Kössel, H. (1993) Editing of the chloroplast rpoB transcript is independent of chloroplast translation and shows different patterns in barley and maize. *EMBO J.* **12**, 4291–4296.

7. Karcher, D. and Bock, R. (1998) Site-selective inhibition of plastid RNA editing by heat shock and antibiotics: a role for plastid translation in RNA editing. *Nucleic Acids Res.* **26**, 1185–1190.

8. Karcher, D. and Bock, R. (2002) Temperature sensitivity of RNA editing and intron splicing reactions in the plastid ndhB transcript. *Curr. Genet.* **41**, 48–52.

9. Kudla, J. and Bock, R. (1999) RNA editing in an untranslated region of the Ginkgo chloroplast genome. *Gene* **234**, 81–86.

10. Kahlau, S., Aspinall, S., Gray, J. C. and Bock, R. (2006) Sequence of the tomato chloroplast DNA and evolutionary comparison of solanaceous plastid genomes. *J. Mol. Evol.* **63**, 194–207.

11. Maier, R. M., Hoch, B., Zeltz, P. and Kössel, H. (1992) Internal editing of the maize chloroplast ndhA transcript restores codons for conserved amino acids. *Plant Cell* **4**, 609–616.

12. Freyer, R., Kiefer-Meyer, M.-C. and Kössel, H. (1997) Occurrence of plastid RNA editing in all major lineages of land plants. *Proc. Natl. Acad. Sci. U S A* **94**, 6285–6290.

13. Fiebig, A., Stegemann, S. and Bock, R. (2004) Rapid evolution of RNA editing sites in a small non-essential plastid gene. *Nucleic Acids Res.* **32**, 3615–3622.

14. Bock, R., Hagemann, R., Kössel, H. and Kudla, J. (1993) Tissue- and stage-specific modulation of RNA editing of the psbF and psbL transcript from spinach plastids – a new regulatory mechanism? *Mol. Gen. Genet.* **240**, 238–244.

15. Karcher, D. and Bock, R. (2002) The amino acid sequence of a plastid protein is developmentally regulated by RNA editing. *J. Biol. Chem.* **277**, 5570–5574.

16. Kahlau, S. and Bock, R. (2008) Plastid transcriptomics and translatomics of tomato fruit development and chloroplast-to-chromoplast differentiation: chromoplast gene expression largely serves the production of a single protein. *Plant Cell* **20**, 856–874.

17. Bock, R., Kössel, H. and Maliga, P. (1994) Introduction of a heterologous editing site into the tobacco plastid genome: the lack of RNA editing leads to a mutant phenotype. *EMBO J.* **13**, 4623–4628.

18. Bock, R., Hermann, M. and Kössel, H. (1996) *In vivo* dissection of *cis*-acting determinants for plastid RNA editing. *EMBO J.* **15**, 5052–5059.

19. Chaudhuri, S. and Maliga, P. (1996) Sequences directing C to U editing of the plastid psbL mRNA are located within a 22 nucleotide segment spanning the editing site. *EMBO J.* **15**, 5958–5964.

20. Bock, R., Hermann, M. and Fuchs, M. (1997) Identification of critical nucleotide positions for plastid RNA editing site recognition. *RNA* **3**, 1194–1200.

21. Hermann, M. and Bock, R. (1999) Transfer of plastid RNA-editing activity to novel sites suggests a critical role for spacing in editing-site recognition. *Proc. Natl. Acad. Sci. U S A* **96**, 4856–4861.

22. Miyamoto, T., Obokata, J. and Sugiura, M. (2002) Recognition of RNA editing sites is directed by unique proteins in chloroplasts: biomedical identification of *cis*-acting elements and *trans*-acting factors involved in RNA editing in tobacco and pea chloroplasts. *Mol. Cell. Biol.* **22**, 6726–6734.

23. Chaudhuri, S., Carrer, H. and Maliga, P. (1995) Site-specific factor involved in the editing of the psbL mRNA in tobacco plastids. *EMBO J.* **14**, 2951–2957.

24. Bock, R. and Koop, H.-U. (1997) Extraplastidic site-specific factors mediate RNA editing in chloroplasts. *EMBO J.* **16**, 3282–3288.

25. Hirose, T. and Sugiura, M. (2001) Involvement of a site-specific *trans*-acting factor and a common RNA-binding protein in the editing of chloroplast mRNAs: development of a chloroplast *in vitro* RNA editing system. *EMBO J.* **20**, 1144–1152.

26. Kotera, E., Tasaka, M. and Shikanai, T. (2005) A pentatricopeptide repeat protein is essential for RNA editing in chloroplasts. *Nature* **433**, 326–330.

27. Okuda, K., Nakamura, T., Sugita, M., Shimizu, T. and Shikanai, T. (2006) A pentatricopeptide repeat protein is a site recognition factor in chloroplast RNA editing. *J. Biol. Chem.* **281**, 37661–37667.

28. Chateigner-Boutin, A.-L., Ramos-Vega, M., Guevara-García, A., Andrés, C., Gutiérrez-Nava, M., Cantero, A., Delannoy, E., Jiménez, L. F., Lurin, C., Small, I. and León, P. (2008) CLB19, a pentatricopeptide repeat protein required for editing of rpoA and clpP chloroplast transcripts. *Plant J.* **56**, 590–602.

29. Bock, R. and Maliga, P. (1995) *In vivo* testing of a tobacco plastid DNA segment for guide RNA function in psbL editing. *Mol. Gen. Genet.* **247**, 439–443.

30. Miyamoto, T., Obokata, J. and Sugiura, M. (2004) A site-specific factor interacts directly with its cognate RNA editing site in chloroplast transcripts. *Proc. Natl. Acad. Sci. U S A* **101**, 48–52.

31. Hegeman, C. E., Hayes, M. L. and Hanson, M. R. (2005) Substrate and cofactor requirements for RNA editing of chloroplast transcripts in *Arabidopsis* in vitro. *Plant J.* **42**, 124–132.

32. Sasaki, T., Yukawa, Y., Wakasugi, T., Yamada, K. and Sugiura, M. (2006) A simple *in vitro* RNA editing assay for chloroplast transcripts using fluorescent dideoxynucleotides: distinct types of sequence elements required for editing of ndh transcripts. *Plant J.* **47**, 802–810.

33. Heller, W. P., Hayes, M. L. and Hanson, M. R. (2008) Cross-competition in editing of chloroplast RNA transcripts *in vitro* implicates sharing of *trans*-factors between different C targets. *J. Biol. Chem.* **283**, 7314–7319.

34. Bock, R. (1998) Analysis of RNA editing in plastids. *Methods* **15**, 75–83.

35. Bock, R. (2004) Studying RNA editing in transgenic chloroplasts of higher plants. *Methods Mol. Biol.* **265**, 345–356.

36. Bock, R. (2001) Transgenic chloroplasts in basic research and plant biotechnology. *J. Mol. Biol.* **312**, 425–438.

37. Bock, R. and Khan, M. S. (2004) Taming plastids for a green future. *Trends Biotechnol.* **22**, 311–318.

38. Maliga, P. (2004) Plastid transformation in higher plants. *Annu. Rev. Plant Biol.* **55**, 289–313.

39. Bock, R. (2007) Plastid biotechnology: prospects for herbicide and insect resistance, metabolic engineering and molecular farming. *Curr. Opin. Biotechnol.* **18**, 100–106.

40. Reed, M. L., Peeters, N. M. and Hanson, M. R. (2001) A single alteration 20 nt 5 to an editing target inhibits chloroplast RNA editing in vivo. *Nucleic Acids Res.* **29**, 1507–1513.

41. Sutton, C. A., Zoubenko, O. V., Hanson, M. R. and Maliga, P. (1995) A plant mitochondrial sequence transcribed in transgenic tobacco chloroplasts is not edited. *Mol. Cell. Biol.* **15**, 1377–1381.

42. Reed, M. L. and Hanson, M. R. (1997) A heterologous maize rpoB editing site is recognized by transgenic tobacco chloroplasts. *Mol. Cell. Biol.* **17**, 6948–6952.

43. Schmitz-Linneweber, C., Tillich, M., Herrmann, R. G. and Maier, R. M. (2001) Heterologous, splicing-dependent RNA editing in chloroplasts: allotetraploidy provides *trans*-factors. *EMBO J.* **20**, 4874–4883.

44. Karcher, D., Kahlau, S. and Bock, R. (2008) Faithful editing of a tomato-specific mRNA editing site in transgenic tobacco chloroplasts. *RNA* **14**, 217–224.

45. Murashige, T. and Skoog, F. (1962) A revised medium for rapid growth and bio assays with tobacco tissue culture. *Physiol. Plant* **15**, 473–497.

46. Svab, Z. and Maliga, P. (1993) High-frequency plastid transformation in tobacco by selection for a chimeric *aadA* gene. *Proc. Natl. Acad. Sci. U S A* **90**, 913–917.

47. Ruf, S., Hermann, M., Berger, I. J., Carrer, H. and Bock, R. (2001) Stable genetic transformation of tomato plastids and expression of a foreign protein in fruit. *Nat. Biotechnol.* **19**, 870–875.

48. Wurbs, D., Ruf, S. and Bock, R. (2007) Contained metabolic engineering in tomatoes by expression of carotenoid biosynthesis genes from the plastid genome. *Plant J.* **49**, 276–288.

49. Dufourmantel, N., Pelissier, B., Garcon, F., Peltier, G., Ferullo, J.-M. and Tissot, G. (2004) Generation of fertile transplastomic soybean. *Plant Mol. Biol.* **55**, 479–489.

50. Dufourmantel, N., Tissot, G., Goutorbe, F., Garcon, F., Muhr, C., Jansens, S., Pelissier, B., Peltier, G. and Dubald, M. (2005) Generation and analysis of soybean plastid transformants expressing *Bacillus thuringiensis* Cry1Ab protoxin. *Plant Mol. Biol.* **58**, 659–668.

51. Altpeter, F., Baisakh, N., Beachy, R., Bock, R., Capell, T., Christou, P., Daniell, H., Datta, K., Datta, S., Dix, P. J., Fauquet, C., Huang, N., Kohli, A., Mooibroek, H., Nicholson, L., Nguyen, T. T., Nugent, G., Raemakers, K., Romano, A., Somers, D. A., Stoger, E., Taylor, N. and Visser, R. (2005) Particle bombardment and the genetic enhancement of crops: myths and realities. *Mol. Breed.* **15**, 305–327.

52. Svab, Z., Hajdukiewicz, P. and Maliga, P. (1990) Stable transformation of plastids in higher plants. *Proc. Natl. Acad. Sci. U S A* **87**, 8526–8530.

53. Golds, T., Maliga, P. and Koop, H.-U. (1993) Stable plastid transformation in PEG-treated protoplasts of *Nicotiana tabacum*. *Biotechnology* **11**, 95–97.

54. O'Neill, C., Horvath, G. V., Horvath, E., Dix, P. J. and Medgyesy, P. (1993) Chloroplast transformation in plants: polyethylene glycol (PEG) treatment of protoplasts is an alternative to biolistic delivery systems. *Plant J.* **3**, 729–738.

55. Hayes, M. L., Reed, M. L., Hegeman, C. E. and Hanson, M. R. (2006) Sequence elements critical for efficient RNA editing of a tobacco chloroplast transcript *in vivo* and in vitro. *Nucleic Acids Res.* **34**, 3742–3754.

56. Chateigner-Boutin, A.-L. and Hanson, M. R. (2002) Cross-competition in transgenic chloroplasts expressing single editing sites reveals shared *cis* elements. *Mol. Cell. Biol.* **22**, 8448–8456.

57. Staub, J. M. and Maliga, P. (1994) Translation of the psbA mRNA is regulated by light via the 5'-untranslated region in tobacco plastids. *Plant J.* **6**, 547–553.

58. Ye, G.-N., Hajdukiewicz, P. T. J., Broyles, D., Rodriguez, D., Xu, C. W., Nehra, N. and Staub, J. M. (2001) Plastid-expressed 5-enolpyruvylshikimate-3-phosphate synthase genes provide high level glyphosate tolerance in tobacco. *Plant J.* **25**, 261–270.

59. Herz, S., Füssl, M., Steiger, S. and Koop, H.-U. (2005) Development of novel types of plastid transformation vectors and evaluation of factors controlling expression. *Transgenic Res.* **14**, 969–982.

60. Zhou, F., Karcher, D. and Bock, R. (2007) Identification of a plastid Intercistronic Expression Element (IEE) facilitating the expression of stable translatable monocistronic mRNAs from operons. *Plant J.* **52**, 961–972.

61. Bohne, A.-V., Ruf, S., Börner, T. and Bock, R. (2007) Faithful transcription initiation from a mitochondrial promoter in transgenic plastids. *Nucleic Acids Res.* **35**, 7256–7266.

62. Oey, M., Lohse, M., Kreikemeyer, B. and Bock, R. (2009) Exhaustion of the chloroplast protein synthesis capacity by massive expression of a highly stable protein antibiotic. *Plant J.* **57**, 436–445.

63. Svab, Z. and Maliga, P. (1991) Mutation proximal to the tRNA binding region of the *Nicotiana* plastid 16S rRNA confers resistance to spectinomycin. *Mol. Gen. Genet.* **228**, 316–319.

64. Rogalski, M., Karcher, D. and Bock, R. (2008) Superwobbling facilitates translation with reduced tRNA sets. *Nat. Struct. Mol. Biol.* **15**, 192–198.

65. Oey, M., Lohse, M., Scharff, L. B., Kreikemeyer, B. and Bock, R. (2009) Plastid production of protein antibiotics against pneumonia via a new strategy for high-level expression of antimicrobial proteins. *Proc. Natl. Acad. Sci. U S A* **106**, 6579–6584.

66. Ayliffe, M. A. and Timmis, J. N. (1992) Tobacco nuclear DNA contains long tracts of homology to chloroplast DNA. *Theor. Appl. Genet.* **85**, 229–238.

67. Stegemann, S., Hartmann, S., Ruf, S. and Bock, R. (2003) High-frequency gene transfer from the chloroplast genome to the nucleus. *Proc. Natl. Acad. Sci. U S A* **100**, 8828–8833.

68. Bock, R. and Timmis, J. N. (2008) Reconstructing evolution: gene transfer from plastids to the nucleus. *Bioessays* **30**, 556–566.

69. Hager, M., Biehler, K., Illerhaus, J., Ruf, S. and Bock, R. (1999) Targeted inactivation of the smallest plastid genome-encoded open reading frame reveals a novel and essential subunit of the cytochrome b6f complex. *EMBO J.* **18**, 5834–5842.

70. Ruf, S., Biehler, K. and Bock, R. (2000) A small chloroplast-encoded protein as a novel architectural component of the light-harvesting antenna. *J. Cell Biol.* **149**, 369–377.

71. Ruf, S., Karcher, D. and Bock, R. (2007) Determining the transgene containment level provided by chloroplast transformation. *Proc. Natl. Acad. Sci. U S A* **104**, 6998–7002.

72. Bock, R. (2007) Structure, function, and inheritance of plastid genomes. *Top. Curr. Genet.* **19**, 29–63.

73. Doyle, J. J. and Doyle, J. L. (1990) Isolation of plant DNA from fresh tissue. *Focus* **12**, 13–15.

74. Peeters, N. M. and Hanson, M. R. (2002) Transcript abundance supercedes editing efficiency as a factor in developmental variation of chloroplast gene expression. *RNA* **8**, 497–511.

75. Timmermans, M. C. P., Maliga, P., Vieira, J. and Messing, J. (1990) The pFF plasmids: cassettes utilising CaMV sequences for expression of foreign genes in plants. *J. Biotechnol.* **14**, 333–344.

Chapter 9

Identifying Specific *Trans*-Factors of RNA Editing in Plant Mitochondria by Multiplex Single Base Extension Typing

Mizuki Takenaka

Abstract

The multiplex single base extension SNP-typing procedure outlined here can be employed to screen large numbers of plants for mutations in nuclear genes that affect mitochondrial RNA editing. The high sensitivity of this method allows high-throughput analysis of individual plants altered in RNA editing at given sites in total cellular cDNA from pooled RNA preparations of up to 50 green plants. The method can be used for large-scale screening for mutations in genes encoding *trans*-factors for specific RNA editing sites. Several nuclear encoded genes involved in RNA editing at specific sites in mitochondria of *Arabidopsis thaliana* have been identified by this approach.

Key words: RNA editing, Plant mitochondria, Mutant screening, SNP typing

1. Introduction

RNA editing in mitochondria of flowering plants changes about 450 specific cytosines to uridines mostly in mRNAs (1, 2). In the past years *cis*-determinants of several RNA editing sites have been characterized by in organello and *in vitro* analyses of RNA editing in mitochondrial RNA molecules, but these studies did not lead to identification of any *trans*-factors (2–5). Only recently have the first of the factors involved in recognition of the specific editing sites been identified (6, 7), while the enzyme(s) involved in the biochemical reaction remain unknown.

The biochemical purification of these *trans*-factors, site-specific or more general, may be difficult if not impossible presumably due to very low concentrations of these proteins. As an alternative and/or complementing approach, screening for mutants provides more rapid access to these factors and their genes. In plastids, nuclear

Ruslan Aphasizhev (ed.), *RNA and DNA Editing: Methods and Protocols*, Methods in Molecular Biology, vol. 718,
DOI 10.1007/978-1-61779-018-8_9, © Springer Science+Business Media, LLC 2011

mutants detected by various secondary phenotypic defects have led to the identification of several genes required for editing at specific sites (8–11). Analogous phenotypic screens are not feasible for mitochondria because most mitochondrial genes code for subunits of the respiratory chain and severe disturbances in any of these genes will usually be lethal. We have therefore developed a method to directly analyse and screen for RNA editing defects at individual sites in populations of randomly mutagenized plants (12). This method combines the multiplexed approaches developed for animal single-nucleotide genotyping (13, 14) with the single base extension procedure employed at plastid editing sites (15).

The procedure can be used to screen large mutant populations of plants for alterations at specific RNA editing sites. This method represents an alternative to a recently developed method based on high-resolution melting (HRM) of amplicons (16). The HRM approach allows the identification of new RNA editing sites and the screening for affected site(s) in an isolated mutant plant line or individual. However, the screening of large mutant populations for a mutant plant affected at a given editing site, which can be done with the multiplexed primer extension protocol outlined below, is not feasible by HRM. The multiplexed base addition approach can, like the HRM, be used to search for the potential RNA editing defects in plants mutated by T-DNA insertion. Candidate genes are some or all of the more than 400 PPR-genes encoded in flowering plant genomes (17, 18).

The multiplexed primer extension procedure can be also employed for identification of nuclear mutations affecting RNA editing in chloroplasts. Screening for mutants of the U to C editing events in plastids and mitochondria of non-flowering plants as well as for mutants of RNA editing in other systems and organisms is possible. However, the subsequent search for the mutant gene will mostly depend on the availability of gene identification procedures. Where these are established in model organisms, such as *Arabidopsis thaliana*, screening will be feasible for any of the various nucleotide differences caused by enzymatic or chemical reactions such as tRNA modifications, rRNA methylation, and even splicing.

2. Materials

2.1. Plant Material

A. thaliana seeds for the wild type Col and C24 ecotypes are required to grow plants for control assays. An EMS-mutant library of *A. thaliana* ecotype Col (and other ecotypes) can be obtained commercially (e.g. Lehle seeds). *A. thaliana* growth conditions are long day for rapid flowering with 16 h light and 8 h dark at 22°C. Further details and conditions can be found in the *Arabidopsis* protocols in this series. Growth time, i.e. age of the

A. thaliana plants should be between 3 and 4 weeks, but is not crucial; age of the leaf is more important.

2.2. Equipment

Access to an automatic sequencer to analyse the labelled single-nucleotide extension products such as an ABI 3100 machine.

2.3. Stock Solutions

1. RNAspin mini kit (GE Healthcare) or any other plant RNA preparation kit.
2. A buffer containing 80 mM Tris–HCl (pH 9.0) and 2 mM $MgCl_2$ to readjust the SNaPshot™ reaction mix from the kit.
3. SNP extension kit ABI SNaPshot™.
4. RT-reaction solutions as recommended by the respective manufacturer of the enzyme. We use M-MLV (Promega) with good results.
5. Alkaline phosphatase (SAP or CIP) with the recommended buffers.

2.4. Primer

1. Design the primers for the RT-PCR step as detailed below in Subheading 3.2 to preferentially access RNA molecules. A good program to design these primers is the DNADynamo. Melting points in 50 mM salt should be in the range of 44–47°C.
2. Design the primers for the specific editing site as oligonucleotides with staggered lengths and with similar melting points as detailed below in Subheading 3.3 and as exemplary given in Table 1. We obtained these primers from Biomers GmbH, Ulm, Germany.

3. Methods

The RNA isolation and handling procedures should be performed at 4°C or on ice with sterile and cold tips, tubes, etc. as described in refs. (19–21).

3.1. Preparation of Total Cellular RNA

1. Collect and pool leaves of similar sizes (1–3 cm in length) from each plantlet to ensure that each individual is represented approximately equally in the RNA preparations.
2. Pool leaves of eight plants into the "small pools" (Fig. 1).
3. Break leaves with a plastic pistil fitting tightly into an Eppendorf tube. Add 350 µl RNA1 buffer from the RNAspin kit (GE Healthcare), mix to dissolve the cell contents and spin in a table top minifuge for 5 min at maximum speed.
4. Combine portions of the first supernatants of the RNA extraction assays from three of the "small pools" of eight plants into "large pools" of 24 plants (Fig. 2; see also note 5).

Table 1
An example of a primer set

Name	Sequence	Length (nt)
>atp1ss1110RA26	AAAAAAAAGCGACTGACAGATAAGCC	26
>nad2ss59FA26	CATGTTCAATCTTTTTTTAGCGGTTT	26
>atp4ss215FA30	AAAAAAAAAATCCAGGCTATTCAGGAAGAAT	30
>atp6ss475RA30	AAAAAAAAAGTTAGCAGCATGAACAAAGAT	30
>atp4ss138RA	AAAAAAAAAAAACCTAAACTCTTACGACTGAATAT	34
>atp4ss248FA34	AAAAAAAAAAAAAACCAATCCTAACGAAGTAGTTC	34
>nad2ss389FA38	AAAAAAAATTGATGCTTTTGAATTCATTGTATTA AGTC	38
>nad4ass124RA	AAAAAAAAAAAAAAAAAAAGTAATAAGAGAGGCA CACA	38
>atp1ss1178FA41	AAAAAAAAAAAAAAAAAAATATGAAACAAGTATGC GGTAGTT	41
>atp4ss89RA41	AAAAAAAAAAAAAAACCTACTATCATTTCTTCATT ATAGATT	41
>ccb206ss137FA44	AAAAAAAAAAAAAAAAAGGTTTTGAAAAAGACTTT TTATGTCATT	44
>nad4ss164RA44	AAAAAAAAAAAAAAAAAAAAAGTAGAAGAGTCGAA TTGTATCAAA	44
>ccb206ss159FA47	AAAAAAAAAAAAAAAAAAAAAAAATTTCCATTTAG GTTTGATTTGGAT	47
>ccb206ss164RA48	AAAAAAAAAAAAAAAAAAAAAAAAAAGTGCAGAAA GAAAAGAAAACAAC	48
>nad2ss89FA51	AAAAAAAAAAAAAAAAAAAAAAAAAAAGAGATCTT TATCATTAATGCAACCT	51
>nad2ss90RA51	AAAAAAAAAAAAAAAAAAAAAAAAAAAATACAACT CCATGAATGAGCAAAAT	51
>nad2ss344FA54	AAAAAAAAAAAAAAAAAAAAAAAAAAAAAATACCAT TTCGATGTGTTTCGATTGTT	54
>nad4ss197RA54	AAAAAAAAAAAAAAAAAAAAAAAAAAAAAAAAAC GAAGGCTTTCCACAAATTGA	54
>atp1ss1292FA57	AAAAAAAAAAAAAAAAAAAAAAAAAAAAAAAAAA AAAAGCAAGGCTGACAGAAGTAC	57
>nad2ss394RA57	AAAAAAAAAAAAAAAAAAAAAAAAAAAAAAAAAA AATAAAGAGCATACCGCAAGTAG	57

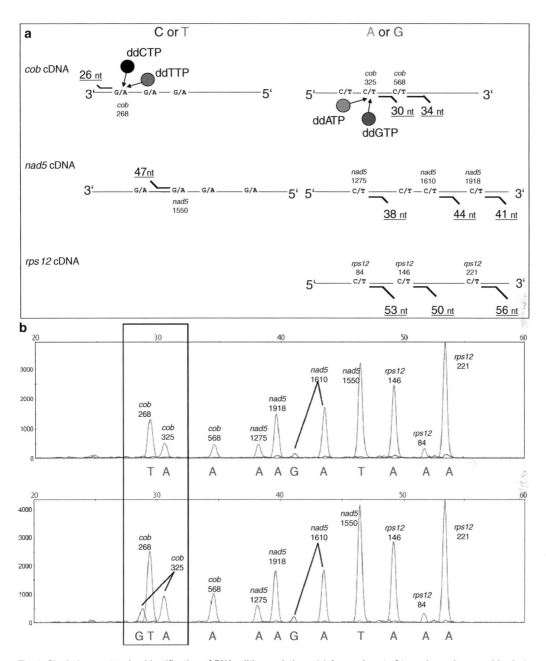

Fig. 1. Single base extension identification of RNA editing variations. (**a**) A sample set of ten primers is spaced by 3–4 nucleotides and designed to address the RNA editing status at ten editing sites. Primer lengths are given by the under-lined numbers in nucleotides (nt). The mitochondrial mRNAs code for cytochrome b (*cob*), subunit 5 of the NADH-dehydrogenase of the respiratory chain (*nad5*) and protein 12 of the small ribosomal subunit (*rps12*). Editing sites can be analysed on either strand of the cDNA, incorporating the ddC/ddT or the ddA/ddG combinations at a monitored editing site. The positions of the editing sites are indicated as the number of nucleotides from the ATG. (**b**) Readout from this primer set shows the spacing of the fluorescence signals obtained by an analysis of wild type *Arabidopsis thaliana* plants (*top part*). Site *nad5*-1610 is not completely edited *in vivo*, the few unedited mRNAs are detected as a G signal. The *bottom part* shows the analysis of a pool of ten plants containing a mutant deficient in RNA editing at site *cob*-325. The mutant is detected by the appearance of a G peak.

c

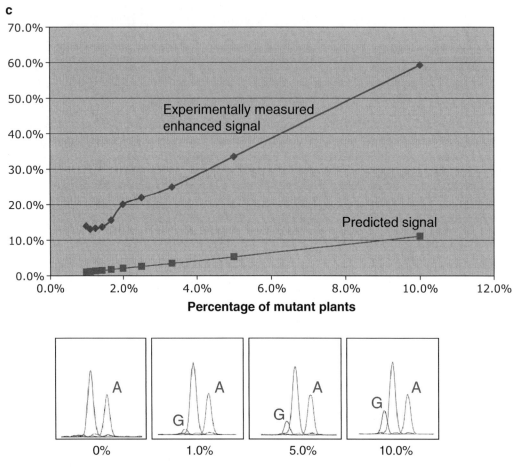

Fig. 1. (continued) (**c**) The signal of the *cob*-325 mutant is used to determine the number of plants which can be pooled. The percentages at the *x*-axis represent the portion of mutant plants, 1% being 1 mutant leaf in 100 leaves. On the *y*-axis, the signal from the unedited nucleotide (seen as either C or G) is shown as the percentage of the signal of the edited nucleotide (read as T or A) for the various percentages of mutant plants assayed. As detailed in the text, the signal is enhanced by limiting the ddNTP pool and increasing the extension cycle number. Comparison of the theoretical and experimental readouts shows this enhancement to be linear above 2–3%. Sample readouts are shown in the *bottom part*. For unambiguous detection about 4–5% mutant plant material are sufficient and for screening 20–25 plants can be pooled (Figure taken from *Nucleic Acids Res* **37,** e13).

Fig. 2. Detection of specific RNA editing mutants in the mutated plant population. (**a**) Green leaves of similar sizes were pooled from eight plants in the "small pools" and three of the small pools were combined for "large pools" of 24 plants. The RT-PCR products from 18 reactions for 16 genes from each "large pool" were analysed for SNPs in six assays. Sample traces of the single base extension assays are shown for pools without any mutants for the monitored RNA editing sites in pool 1. In pool 2, a mutant is identified which has lost editing at site *nad2*-842 and pool 8 contains a mutation of editing at site *cox3*-422. The gene *nad2* codes for subunit 2 of the NADH-dehydrogenase of the respiratory chain and *cox3* codes for subunit 3 of the cytochrome oxidase of the respiratory chain. (**b**) The mutated individual plants are identified in the vertical and horizontal pools of eight plants each. (**c**) The isolated individual plants are compared with wild type primer extensions for the editing levels at sites *nad2*-842 (*line 5-C*) and *cox3*-422 (*line 24-E*). No trace of an edited nucleotide is detected, showing that the underlying mutations have incapacitated genes required for editing at these respective sites (Figure taken from *Nucleic Acids Res* **37,** e13).

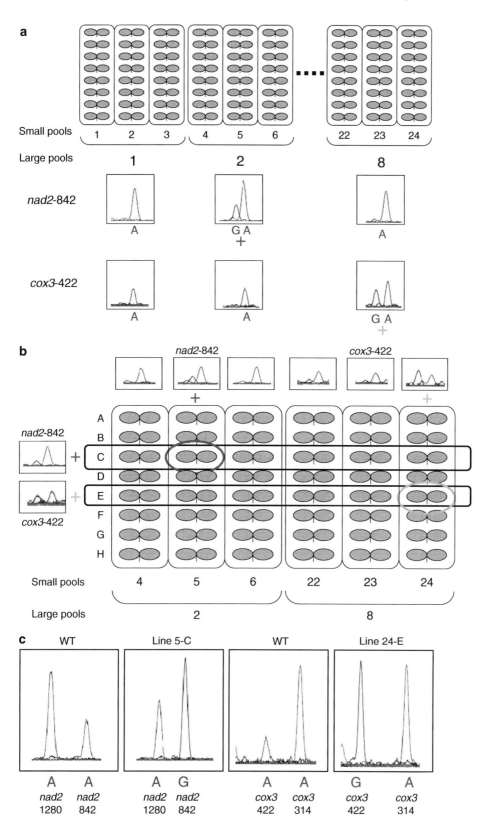

5. Further treatment of the RNA samples follows the protocol from the RNAspin mini kit (GE Healthcare).

3.2. Preparation of cDNA Fragments

1. Design downstream primers for first-strand synthesis to terminate at an RNA editing site downstream of the one you want to assay (see Note 1). RNA editing sites can be found in the pertinent database (22). The last nucleotide (the 3′ terminal nucleotide) should be an A to anneal only at edited mRNA. To increase the effect of this terminal mismatch to genomic unedited DNA you should introduce an artificial mismatch two nucleotides downstream of the editing site. To further lower the probability of the primer binding to residual genomic mitochondrial DNA, the primer should cover one or more other editing site(s) if possible and again should correspond only to the edited mRNA. The presence of an intron can be helpful: if located within the amplified fragment, the intron would yield a large or no product from the genomic DNA. A rapid gel (like 1% agarose gel in TBE) analysis can be used to assess the purity of the amplified cDNA. One of the primers can also be designed to cover an exon–exon border. Generate gene-specific cDNA fragments in independent reactions for each fragment (see Note 2).

2. Synthesize the first strand by a 1-min RT-extension at an appropriate annealing temperature. This standard RT should be sufficient for up to 600 nt of first strand and should not copy large intron-containing DNA fragments.

3. Amplify the first cDNA strand by PCR under standard conditions for 35 cycles (35 times: 45 s 95°C, 45 s 53°C, 1 min 72°C).

4. Clean up PCR products by treatment with 1 U SAP or CIP and by removing the primers with 1 U Exo 1 for 60 min at 37°C and inactivate the enzymes by 15 min at 75°C as described in the SNaPshot protocol.

3.3. Preparation of Primers for Multiplex SNaPshot Analysis

For the multiplex SNP assays, design specific oligonucleotides (e.g. Biomers, Ulm, Germany) with regard to staggered lengths of 3–5 nucleotides and to similar annealing/melting temperatures in the separately mixed batches of 10–20 such primers. Ensure that the primers do not anneal to each other or fold into stable secondary structures as detailed in the SNaPshot protocol instructions.

An example of such a primer set is given in Table 1.

3.4. Multiplex SNP Analysis of the cDNAs

1. Pool about 5–10 ng of each RT-PCR fragment for the separate single-nucleotide primer extension analyses (see Note 3). The amount of PCR product visible as a band on a gel (see Subheading 3.2, step 1) is fine. Different relative amounts of

the various RT-PCR products from different mRNAs are tolerated.

2. We routinely combine 10–20 primers in a single reaction. If possible select your primer combinations so that several of the targeted editing sites are located on a single RT-PCR product. This reduces the number of fragments you have to mix in a given SNaPshot assay. Use 0.6 pmol of each primer in the extension reactions.

3. For the single-nucleotide primer extension reactions follow the instructions of the commercial kits (e.g. ABI SNaPshot™), but reduce the concentrations of the SNaPshot mix to 1/12th of the recommended amounts for optimal enhancement of the unedited C-signal. To maintain the necessary magnesium concentration in the routinely performed 5 µl half-size reaction volumes, add 2.1 µl of a buffer containing 80 mM Tris–HCl (pH 9.0), and 2 mM $MgCl_2$ to the 0.4 µl reaction mix.

4. Increase the single-nucleotide primer extension cycle numbers to 60 cycles to further enhance the unedited mutant C-signal after depletion of the edited wild type T-corresponding nucleotide. To achieve this, the SNaPshot mix was diluted (Subheading 3.4, step 3).

5. Analyse samples on an automated sequencing machine such as the ABI 3100 (see Notes 4 and 5).

4. Notes

1. Amplified fragments should be designed to cover several editing sites to allow parallel screening with the SNP primers.

2. Simultaneous amplification of several fragments in one reaction is possible, depending on the primer combinations. Mostly, such pooled reactions proved in our hands to be too variable to yield comparable amounts of products, but should be worth testing for the fragments you want to produce.

3. Test out the optimal template amounts. Too little of a given template PCR fragment will result in an increase of unspecific peaks in the background. Generally the 5–10 ng of a PCR fragment visible as a band on a gel will be sufficient.

4. Usually the C and G peaks run faster than the T and A peaks. Furthermore, the distances between the C and T or the G and A peaks depend on the primer compositions and are thus different for each primer.

5. Include a positive control, if possible, and test for the observed peak height. If the signal peak is too low, you may have to

reduce the number of plants in a given pool. For example, if the signal obtained from 1 in 24 plants (8×3 plants) is too low, you should try with a pool of only 18 plants (6×3 plants).

Acknowledgements

I thank Dagmar Pruchner for excellent experimental help. I am very grateful to Walther Vogel, Christiane Maier and Bärbel Weber in the Institut für Humangenetik, Universität Ulm, for the use of the automated sequencing equipment and materials and their very constructive and helpful discussions throughout. This work was supported by grants from the DFG.

References

1. Giegé, P., and Brennicke, A. (1999) RNA editing in *Arabidopsis* mitochondria effects 441 C to U changes in ORFs. *Proc Natl Acad Sci U S A* **96**, 15324–15329.

2. Takenaka, M., Verbitskiy, D., van der Merwe, J. A., Zehrmann, A., and Brennicke, A. (2008) The process of RNA editing in plant mitochondria. *Mitochondrion* **8**, 35–46.

3. Farré, J.-C., Leon, G., Jordana, X., and Araya, A. (2001) *cis* Recognition elements in plant mitochondrion RNA editing. *Mol Cell Biol* **21**, 6731–6737.

4. Neuwirt, J., Takenaka, M., van der Merwe, J. A., and Brennicke, A. (2005) An *in vitro* RNA editing system from cauliflower mitochondria: editing site recognition parameters can vary in different plant species. *RNA* **11**, 1563–1570.

5. van der Merwe, J. A., Takenaka, M., Neuwirt, J., Verbitskiy, D., and Brennicke, A. (2006) RNA editing sites in plant mitochondria can share *cis*-elements. *FEBS Lett* **580**, 268–272.

6. Zehrmann, A., Verbitskiy, D., van der Merwe, J. A., Brennicke, A., and Takenaka, M. (2009) A DYW-domain containing PPR-protein is required for RNA editing at multiple sites in mitochondria of *Arabidopsis thaliana*. *Plant Cell* **21**, 558–567.

7. Zehrmann, A., van der Merwe, J. A., Verbitskiy, D., Brennicke, A., and Takenaka, M. (2008) Ecotype-specific variations in the extent of RNA editing in plant mitochondria. *Mitochondrion* **8**, 319–327.

8. Kotera, E., Tasaka, M., and Shikanai, T. (2005) A pentatricopeptide repeat protein is essential for RNA editing in chloroplasts. *Nature* **433**, 326–330.

9. Okuda, K., Myouga, R., Motohashi, K., Shinozaki, K., and Shikanai, T. (2007) Conserved domain structure of pentatricopeptide repeat proteins involved in chloroplast RNA editing. *Proc Natl Acad Sci U S A* **104**, 8178–8183.

10. Okuda, K., Nakamura, T., Sugita, M., Shimizu, T., and Shikanai, T. (2006) A pentatricopeptide repeat protein is a site-recognition factor in chloroplast RNA editing. *J Biol Chem* **281**, 37661–37667.

11. Chateigner-Boutin, A.-L., Ramos-Vega, M., Guevara-Garcia, A., Andrés, C., Gutierrez-Nava, M. d. l. L., Cantero, A., Delannoy, E., Jimenez, L. F., Lurin, C., Small, I. D., and León, P. (2008) CLB19, a pentatricopeptide repeat protein required for editing of rpoA and clpP chloroplast transcripts. *Plant J* **56**, 590–602.

12. Takenaka, M., and Brennicke, A. (2008) Multiplex single base extension typing to identify nuclear genes required for RNA editing in plant organelles. *Nucleic Acids Res.* doi: 10.1093/nar/gkn975

13. Lindblad-Toh, K., Winchester, E., Daly, M. J., Wang, D. G., Hirschhorn, J. N., Laviolette, J.-P., Ardlie, K., Reich, D. E., Robinson, E., Sklar, P., Shah, N., Thomas, D., Fan, J. B., Gingeras, T., Warrington, J., Patil, N., Hudson, T. J., and Lander, E. S. (2000) Large-scale discovery and genotyping of single-nucleotide polymorphisms in the mouse. *Nat Genet* **24**, 381–386.

14. Nelson, T. M., Just, R. S., Loreille, O., Schanfield, M. S., and Podini, S. (2007) Development of a multiplex single base extension assay for mitochondrial DNA haplotype typing. *Croat Med J* **48**, 460–472.

15. Sasaki, T., Yukawa, Y., Wakasugi, T., Yamada, K., and Sugiura, M. (2006) A simple *in vitro* RNA editing assay for chloroplast transcripts using fluorescent dideoxynucleotides: distinct types of sequence elements required for editing of ndh transcripts. *Plant J* **47**, 802–810.

16. Chateigner-Boutin, A.-L., and Small, I. D. (2007) A rapid high-throughput method for the detection and quantification of RNA editing based on high-resolution melting of amplicons. *Nucleic Acids Res* **35**, e114.

17. Lurin, C., Andrés, C., Aubourg, S., Bellaoui, M., Bitton, F., Bruyère, C., Caboche, M., Debast, C., Gualberto, J. M., Hoffmann, B., Lecharny, A., Le Ret, M., Martin-Magniette, M.-L., Mireau, H., Peeters, N., Renou, J.-P., Szurek, B., Taconnat, L., and Small, I. D. (2004) Genome-wide analysis of *Arabidopsis* pentatricopeptide repeat proteins reveals their essential role in organelle biogenesis. *Plant Cell* **16**, 2089–2103.

18. Andrés, C., Lurin, C., and Small, I. D. (2007) The multifarious roles of PPR proteins in plant mitochondrial gene expression. *Physiol Plant* **129**, 14–22.

19. Takenaka, M., and Brennicke, A. (2007) RNA editing in plant mitochondria: assays and biochemical approaches. *Methods Enzymol* **424**, 439–458.

20. *Arabidopsis Protocols* (Martínez-Zapater, J. M., and Salinas, J. eds.), (1998) Methods in Molecular Biology 82, Humana, Totowa, NJ.

21. *Plant Gene Transfer and Expression Protocols* (Jones, H. ed.), (1995) Methods in Molecular Biology 49, Humana, Totowa, NJ.

22. Picardi, E., Regina, T. M. R., Brennicke, A., and Quagliariello, C. (2007) REDIdb: the RNA editing database. *Nucleic Acids Res* **35**, D173–D177.

Chapter 10

Complementation of Mutants in Plant Mitochondrial RNA Editing by Protoplast Transfection

Mizuki Takenaka and Anja Zehrmann

Abstract

A crucial and often decisive test of a nuclear gene being involved in a given process is the complementation of mutants. Restoring the wild type phenotype by the wild type gene introduced into the mutant is a major piece of evidence for the function of this gene. We have developed a rapid and reliable method to complement protoplasts from plants with mutations in mitochondrial RNA editing with the respective wild type genes. The method furthermore allows testing the functionality of modified protein sequences without the need to make and grow transgenic plants, which is very time-consuming. We successfully employed this method for several nuclear-encoded genes involved in RNA editing at specific sites in mitochondria of *Arabidopsis thaliana*.

Key words: RNA editing, Plant mitochondria, Mutant screening, SNP typing

1. Introduction

RNA editing in mitochondria of flowering plants changes several hundred specific cytidines to uridines (1, 2). In the past years, *cis*-determinants of several RNA editing sites have been characterized by in organello and *in vitro* analyses of RNA editing in mitochondrial RNA molecules (2–5). Recently, the first *trans*-factor involved in recognition of specific editing sites in plant mitochondria has been identified (6, 7), while the enzyme(s) involved in the biochemical reaction is(are) still unknown. This editing protein as well as several editing proteins described for specific editing events in plastids (8–10) – so far all of them pentatricopeptide repeat (PPR) proteins – have been identified through phenotypic variations in editing to the nuclear genomic loci involved. In such investigations, sequence analysis and comparison of the mutant

Ruslan Aphasizhev (ed.), *RNA and DNA Editing: Methods and Protocols*, Methods in Molecular Biology, vol. 718,
DOI 10.1007/978-1-61779-018-8_10, © Springer Science+Business Media, LLC 2011

allele with the wild type allele will strongly suggest that the gene identified is indeed involved in this RNA editing event.

However, to verify the crucial requirement of these site-specific *trans*-factors, the mutants through which they were identified will ultimately have to be complemented. This complementation can be done by stable transformation of the mutant plants, which is very time-consuming since it requires growing the transformed plants, seed set, and analysis of at least the F1-generation.

We have now adapted a procedure for transient transfection of protoplasts (11) from various wild type ecotypes and mutant lines of *Arabidopsis thaliana*, and used this approach to complement the phenotypic defects in RNA editing in plant mitochondria (6, 7). The successful transfection of protoplasts suggests that the introduced genes are correctly transcribed and translated, that the protein products are effectively imported into mitochondria and that they are functionally assembled into the RNA editing machinery. Most likely this procedure can also be employed to investigate RNA editing factors in plastids of *A. thaliana*, since the protoplasts are prepared from young green leaves.

2. Materials

2.1. Plant Material

A. thaliana seeds for the wild type Columbia (Col; or the respective other ecotype in which the mutants were made) are required to raise plants for control assays. Grow wild type and mutant lines of *A. thaliana* under growth conditions according to the standards as can be found in the *Arabidopsis* protocols in this series (12, 13). Growth time, i.e., age of the *A. thaliana* plants is not crucial, age of the leaf is more important.

2.2. Equipment

Access to a fluorescence microscope with filters for GFP and chloroplast autofluorescence is necessary. Nucleotide sequence analyses can be obtained commercially.

2.3. Stock Solutions

1. 0.2 M 4-Morpholineethanesulfonic acid (MES, pH 5.7), sterilized by autoclaving.
2. 0.8 M Mannitol, sterilized by filtration.
3. 1 M $CaCl_2$, sterilized by filtration.
4. 2 M KCl, sterilized by filtration.
5. 2 M $MgCl_2$, sterilized by filtration.
6. β-Mercaptoethanol.
7. 10% (wt/vol) BSA, sterilized by filtration.
8. Cellulase R10 (e.g., Yakult Pharmaceutical Ind. Co., Ltd., Japan).

9. Macerozyme R10 (e.g., Yakult Pharmaceutical Ind. Co., Ltd., Japan).

10. PEG4000 (very important to use a good supplier, e.g., Fluka).

11. 1 M Tris–phosphate (pH 7.8), sterilized by autoclaving.

12. 2 mM MES (pH 5.7), 154 mM NaCl, 125 mM CaCl$_2$, and 5 mM KCl.

Derivatives of these and additional solutions are given in the next section and at the appropriate steps.

2.4. Fresh Solutions

1. Prepare the enzyme solution freshly before use containing 20 mM MES (pH 5.7), 1.5% (wt/vol) cellulase R10, 0.4% (wt/vol) macerozyme R10, 0.4 M mannitol, and 20 mM KCl. Heat the solution to 55°C for 10 min to inactivate DNase and proteases and to improve the enzyme solubility. Cool to room temperature and add 10 mM CaCl$_2$, 1 mM β-mercaptoethanol, and 0.1% BSA. All solutions, especially stocks containing mannitol, should be filtered through 0.45-μm mesh filters (Cellulose acetate; VWR).

2. The transfection solution containing 40% (wt/vol) PEG4000, 0.2 M mannitol, and 100 mM CaCl$_2$ should be made at least 1 h before transfection to completely dissolve the PEG, but should be used within 3 days.

3. Methods

3.1. Preparation of Protoplasts

The procedure closely follows the method as described in (11).

1. Collect and pool leaves of similar sizes (1–3 cm in length) from 3 to 4 weeks old healthy unstressed plantlets.

2. Cut 0.5- to 1-mm leaf strips from the center part of a leaf with a new razor blade without damaging the cells around the cut site. For routine experiments, 10–20 leaves digested in 5–10 ml enzyme solution will give 0.5–1 × 10^6 protoplasts, enough for about 25–100 assays.

3. Transfer leaf strips quickly into the freshly prepared enzyme solution (10–20 leaves in 5–10 ml) by completely submerging the leaf strips. Do not let the leaf slices dry out.

4. Vacuum infiltrate leaf strips for 30 min in the dark in a desiccator.

5. Continue the enzymatic digestion, without shaking, in the dark for at least 3 h at room temperature. The enzyme solution will turn green after gently swirling, indicating the release of the protoplasts. Digestion time should be optimized for each ecotype and genotype, although 4 h are usually optimal.

6. Control the release of protoplasts in the solution under the microscope, protoplasts should be visible as rounded shapes.

7. Dilute the enzyme/protoplast solution with an equal volume of a solution containing 2 mM MES (pH 5.7), 154 mM NaCl, 125 mM $CaCl_2$, and 5 mM KCl before filtration through a clean miracloth to remove undigested leaf tissues.

8. Centrifuge the flow-through at $100 \times g$ to pellet the protoplasts for 1–2 min and remove as much supernatant as possible.

9. Resuspend protoplasts in a solution containing 2 mM MES (pH 5.7), 154 mM NaCl, 125 mM $CaCl_2$, and 5 mM KCl at a concentration of about 4×10^5/ml. For counting the protoplasts, the easiest is to use a cell counter slide for the microscope. Keep the protoplasts on ice for 30 min.

10. Protoplasts begin to settle at the bottom of the tube after 15 min. Remove as much buffer as possible without touching the protoplast pellet. Resuspend protoplasts at 4×10^5/ml in a buffer containing 4 mM MES (pH 5.7), 0.4 M mannitol, and 15 mM $MgCl_2$ at room temperature.

11. Add 30 μl DNA (30 μg of plasmid DNA of 5–10 kb in size) to a 2-ml microfuge tube.

12. The plasmid should be clean without contaminants such as in preparations obtained with, e.g., the purelink Hipure plasmid maxi kit (Invitrogen).

13. Add 300 μl protoplasts (1.2×10^5 protoplasts) and mix gently.

14. Add 330 μl of a solution containing 40% (wt/vol) PEG4000, 0.2 M mannitol, and 100 mM $CaCl_2$ at room temperature. This solution should be made at least 1 h before transfection to completely dissolve the PEG. Mix solutions gently.

15. Incubate the mixture at room temperature for 5–15 min.

16. Dilute the transfection assay with 1.2 ml of a solution containing 2 mM MES (pH 5.7), 154 mM NaCl, 125 mM $CaCl_2$, and 5 mM KCl at room temperature and mix well to terminate the transfection process.

17. Centrifuge at $200 \times g$ for 2 min at room temperature and remove supernatant.

18. Resuspend protoplasts gently with 1.5 ml of a solution containing 4 mM MES (pH 5.7), 0.5 M mannitol, and 20 mM KCl at room temperature in, e.g., a 6-cm tissue culture dish (Sarstedt).

19. Incubate protoplasts at room temperature (20–25°C) for the time required to express the introduced gene. Usually an overnight (14–24 h) incubation time is sufficient to obtain a clear change in the RNA editing phenotype, but sometimes longer incubation may be necessary. Do not let the protoplasts dry out during longer incubation times by sealing the

plates with parafilm. Be aware that the protoplasts will begin to regenerate the cell walls upon longer incubation.

20. Resuspend and harvest protoplasts by centrifugation at $200 \times g$ for 2 min.

21. Remove the supernatant and shock-freeze samples in liquid nitrogen. Samples can be stored at −80°C until further analysis.

22. Prepare RNA (or DNA for control) from protoplast samples by standard methods, e.g., (12, 13).

23. Design downstream primers for first-strand synthesis to terminate at an RNA editing site downstream of the one you want to assay. RNA editing sites can be found in the pertinent database (14). The last nucleotide (the 3′ terminal nucleotide) should be an A to anneal only at edited mRNA. The procedure to generate respective cDNA fragments and their analysis are detailed in Chapter 9.

4. Notes

1. Vectors for transfection. The main requirement of the introduced vector is a CaMV promoter and the standard transcription, translation, and termination signals in the appropriate positions relative to the gene assayed.

2. Controls. Efficiency of the protoplast transfection can be monitored by cotransfection of a vector containing the GFP gene controlled by the same CaMV promoter. We routinely observe 80–90% of the protoplasts to show GFP fluorescence, i.e., to be transfected.

3. For negative controls only this GFP-vector can be introduced into the protoplasts from the mutant line. Untransfected wild type protoplasts should also be assayed since we often found less effective *in vivo* editing in protoplast preparations than in whole leaf assays. If you want to visualize the GFP fluorescence, it is better to store the protoplasts in the dark.

4. Mitochondrial import assays. If the investigated gene is cloned in frame upstream of the GFP coding sequence, the localization of the fluorescence in the cell can be used to deduce the probable location of the assayed protein. Furthermore, RNA editing complementation can be investigated also in these assays which will allow conclusions as to whether the added GFP will inhibit the RNA editing process in which this particular investigated protein is involved.

5. Editing efficiency in protoplast mitochondria. RNA editing efficiency at the investigated site may in protoplasts be different

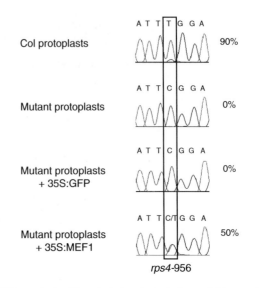

Col protoplasts — A T T [T] G G A — 90%

Mutant protoplasts — A T T [C] G G A — 0%

Mutant protoplasts + 35S:GFP — A T T [C] G G A — 0%

Mutant protoplasts + 35S:MEF1 — A T T [C/T] G G A — 50%

rps4-956

Fig. 1. RNA editing analysis of transfected protoplasts from wild type and mutant plant lines. Untransfected protoplasts from the wild type Col plants show nearly complete RNA editing at the *rps4*-956 site, slightly less than the complete C to U conversion of wild type Col leaves (*top part* and data not shown). A plant with a mutation in the *AtMEF1* gene (6, 7) shows no editing at this *rps4*-956 site in protoplasts or leaves (*second panel from top* and data not shown). Control transfection with the GFP gene does not alter the absence of detectable editing in the mutant protoplasts (35S:GFP). Transfection of protoplasts from the mutant plant with the wild type Col *AtMEF1* gene restores the ability to edit this site in the *rps4* mRNAs to an editing level of about 50% (35S:MEF1).

from the in planta editing (Fig. 1). It is therefore necessary to compare editing also in untransfected mutant and wild type protoplasts with editing in leaves to correctly evaluate the effect of the introduced gene.

Acknowledgments

We thank Dagmar Pruchner for excellent experimental help. This work was supported by a grant from the DFG to M.T.

References

1. Giegé, P., and Brennicke, A. (1999) RNA editing in Arabidopsis mitochondria effects 441 C to U changes in ORFs. *Proc Natl Acad Sci USA* **96,** 15324–15329.

2. Takenaka, M., Verbitskiy, D., van der Merwe, J. A., Zehrmann, A., and Brennicke, A. (2008) The process of RNA editing in plant mito-chondria. *Mitochondrion* **8,** 35–46.

3. Farré, J.-C., Leon, G., Jordana, X., and Araya, A. (2001) Cis recognition elements in plant

mitochondrion RNA editing. *Mol Cell Biol* **21,** 6731–6737.

4. Neuwirt, J., Takenaka, M., van der Merwe, J. A., and Brennicke, A. (2005) An *in vitro* RNA editing system from cauliflower mito-chondria: editing site recognition parameters can vary in different plant species. *RNA* **11,** 1563–1570.

5. van der Merwe, J. A., Takenaka, M., Neuwirt, J., Verbitskiy, D., and Brennicke, A. (2006)

RNA editing sites in plant mitochondria can share cis-elements. *FEBS Lett* **580,** 268–272.

6. Zehrmann, A., Verbitskiy, D., van der Merwe, J. A., Brennicke, A., Takenaka, M. (2009) A DYW-domain containing PPR-protein is required for RNA editing at multiple sites in mitochondria of *Arabidopsis thaliana. Plant Cell* **21,** 558–567.

7. Zehrmann, A., van der Merwe, J. A., Verbitskiy, D., Brennicke, A., and Takenaka, M. (2008) Ecotype-specific variations in the extent of RNA editing in plant mitochondria. *Mitochondrion* **8,** 319–327.

8. Kotera, E., Tasaka, M., and Shikanai, T. (2005) A pentatricopeptide repeat protein is essential for RNA editing in chloroplasts. *Nature* **433,** 326–330.

9. Okuda, K., Myouga, R., Motohashi, K., Shinozaki, K., and Shikanai, T. (2006) Conserved domain structure of pentatricopeptide repeat proteins involved in chloroplast RNA editing. *Proc Natl Acad Sci USA* **104,** 8178–1883.

10. Chateigner-Boutin, A.-L., Ramos-Vega, M., Guevara-Garcia, A., Andrés, C., Gutierrez-Nava, MdlL., Cantero, A., Delannoy, E., Jimenez, L. F., Lurin, C., Small, I. D., and León, P. (2008) CLB19, a pentatricopeptide repeat protein required for editing of rpoA and clpP chloroplast transcripts. *Plant J* **56,** 590–602.

11. Yoo, S.-D., Cho, Y.-H., and Sheen, J. (2007) *Arabidopsis* mesophyll protoplasts: a versatile cell system for transient gene expression analysis. *Nat Protoc* **2,** 1565–1572.

12. *Arabidopsis Protocols* (Martínez-Zapater, J. M., and Salinas, J. eds), (1998) Methods in Molecular Biology 82, Humana, Totowa, NJ.

13. *Plant Gene Transfer and Expression Protocols* (Jones, H. ed.), (1995) Methods in Molecular Biology 49, Humana, Totowa, NJ.

14. Picardi, E., Regina, T. M. R., Brennicke, A., and Quagliariello, C. (2007) REDIdb: the RNA editing database. *Nucleic Acids Res* **35,** D173–D177.

Chapter 11

A High-Throughput Assay for DNA Deaminases

Meng Wang, Cristina Rada, and Michael S. Neuberger

Abstract

Most members of the AID/APOBEC family of polynucleotide deaminases can catalyse the deamination of cytosine to uracil in DNA. They thereby function as active DNA mutators. Here, we describe how bacterial papillation assays can be adapted to monitor the mutator activity of AID/APOBEC proteins and show how such papillation assays can be used as a high-throughput screen to identify AID variants with increased specific activity. It should also be possible to use papillation assays for the identification of novel DNA deaminases.

Key words: Papillation assay, Activation-induced deaminase, APOBEC, DNA deaminase, Cytosine deamination

1. Introduction

Members of the AID/APOBECs family [activation-induced deaminase (AID), apolipoprotein B mRNA editing enzyme catalytic subunit 1 (APOBEC1), APOBEC-like protein 3 (APOBEC3)] are able to deaminate deoxycytidine to deoxyuridine in DNA, which if left unrepaired, will result following replication in transition mutations at C:G pairs.

In the adaptive immune system, AID-triggered deamination of deoxycytidine residues in the immunoglobulin loci acts as a trigger for antibody gene diversification by somatic hypermutation or gene conversion in the immunoglobulin variable region genes, or by IgM→IgG/IgA/IgE class-switch recombination following deamination upstream of the constant region genes (1). The APOBEC3 proteins act as part of the innate immune system restricting invading nucleic acid and retroviral infection: here, for example, APOBEC3F and APOBEC3G deaminate

Ruslan Aphasizhev (ed.), *RNA and DNA Editing: Methods and Protocols*, Methods in Molecular Biology, vol. 718,
DOI 10.1007/978-1-61779-018-8_11, © Springer Science+Business Media, LLC 2011

deoxycytidine residues in the HIV-1 first-strand cDNA replication intermediate (2). APOBEC1, the founder member of the AID/APOBEC family, acts on cytidine in RNA where it edits the mRNA of apolipoprotein B by deaminating cytosine 6666 to uracil to create a premature stop codon. This results in a shortened protein that is then utilized in the synthesis of chylomicrons (3, 4). However, although the first member of the AID/APOBEC family to be identified, APOBEC1 is a later evolutionary arrival than AID from which it likely arose through gene duplication (5). APOBEC1 also retains the ability to deaminate deoxycytidine in DNA (6) although it is unknown as yet whether such DNA deamination by APOBEC1 does (like that by AID and APOBEC3s) fulfil a physiological function.

Many of the techniques currently used to study the activity of AID/APOBEC DNA deaminases rely on the detection of the deoxyuridine generated via deamination. Thus, biochemical assays often utilize a fluorescein-conjugated oligonucleotide that is cleaved to a shorter fragment upon incubation with uracil-DNA glycosylase and alkali following deamination (7), or use a plasmid which following mutation through deamination confers a detectable phenotype when transformed into bacteria, for example, a Lac⁻ phenotype or antibiotic resistance (8–11). A major limiting factor in these assays is that they require at least partial purification of the recombinant DNA deaminases (which are usually poor in expression and solubility), rendering them unsuitable for high-throughput screens.

In vivo assays have also been used which rely on the expression of DNA deaminases in bacteria to induce a mutator phenotype, with these assays obviously not requiring the deaminase investigated to be purified. Mutations introduced by the DNA deaminases can be quantified by measuring the number of cells that are able to grow under selection (12, 13). Several target genes in *Escherichia coli*, when mutated, are able to generate a positive selectable marker. For example, the rifampicin assay relies on mutations in the *rpoB* gene coding for the beta subunit of RNA polymerase giving rise to rifampicin antibiotic resistance (7, 12, 13). Other genomic targets have also been used: the *sacB* gene (mutations give sucrose resistance) and *kanL94P* (C to U mutation restores resistance to kanamycin) (14, 15).

A major limiting factor in all the assays discussed above is that they are low-throughput and thus not suitable for carrying out large-scale mutagenesis studies of DNA deaminases. The papillation assay has been a widely used technique for the isolation of bacterial colonies exhibiting high frequency of mutation (16–21). Here, we describe how the papillation assay can be adapted as high-throughput assay for the DNA deamination activity of members of the AID/APOBEC family.

2. Materials

2.1. Preparing MacConkey-Lactose Agar

1. MacConkey-lactose dehydrated culture media (BD Biosciences).
2. 1 L sterile water.
3. 2-L sterile conical flask.
4. Water bath set to 50°C.
5. 245 mm × 245 mm × 25 mm large square bioassay dishes or 145-mm diameter round petri dishes (NUNC).
6. 100 mg/ml ampicillin chloride dissolved in sterile water stored at −20°C.
7. 1 M Isopropyl β-ᴅ-1-thiogalactopyranoside (IPTG) dissolved in sterile water stored at −20°C.

2.2. Papillation Assay to Detect AID/ APOBECs

1. *E. coli* K12 strain CC102 *araAΔ*(*gpt-lac*)5 carrying F′ *lacproA⁺*, B⁺ episome in which the *lacI* carries a + 1 frameshift mutation, and the *lacZ* carries a GAG → GGG missense mutation at codon 461.
2. AID/APOBEC cDNA cloned into bacterial expression vector pTrc99a (carries ampicillin resistance gene, *lacI*, and the trc promoter to control the expression of AID/APOBEC).
3. MacConkey-lactose agar.
4. Minimal glucose agar (M9 + 0.2% glucose agar).
5. M9 salt solution (70 g $Na_2HPO_4 \cdot 7H_2O$, 30 g KH_2PO_4, 5 g NaCl, 10 g NH_4Cl in 1 L sterile water to give 10× stock).
6. 2× TY media.
7. Water bath set to 42°C.
8. 37°C incubator.
9. Primers to sequence the relevant region of *lacZ* encompassing codon 461 (forward: 5′-AGAATTCCTGAAGTTCAGATGT and reverse: 5′-GGAATTCGAAACCGCCAAGAC).

2.3. Selection for AID/ APOBEC Upmutants

1. Wild-type AID cDNA.
2. BIOTAQ™ DNA polymerase (Bioline).
3. 10× Reaction Buffer (Bioline): 160 mM $(NH_4)_2SO_4$, 670 mM Tris–HCl (pH 8.8 at 25°C), 0.1% stabilizer.
4. 25 mM of each dNTPs in TE buffer (10 mM Tris, 1 mM EDTA, pH 8.0).
5. 20 mM $MgCl_2$.
6. 10 μM primers in TE buffer (forward: 5′-ATGGAATTC-ATGGACAGCCTCTTG and reverse: 5′-CTGAAGCTTT-CAAAGTCCCAAAGTA).

7. Sterile water.

8. PCR thermocycler.

9. *Eco*RI restriction endonuclease (NEB).

10. *Hin*dIII restriction endonuclease (NEB).

11. Qiagen QIAquick PCR Purification Kit.

12. Qiagen Gel Extraction Kit.

13. Qiagen HiSpeed Maxi Kit.

14. Qiagen QIAprep Spin Miniprep Kit.

15. Rapid Ligation Kit (Roche).

16. Competent DH5α *E. coli*.

17. LB agar supplemented with 100 µg/ml ampicillin.

18. Primers to sequence the cloned inserts in pTrc99a plasmid (forward: 5′-GTGTGGAATTGTGAGCGGA and reverse: 5′-GCTGAAAATCTTCTCTCATCC).

19. All materials from Subheading 2.2.

2.4. Using Arabinose Induction to Select for AID/APOBEC Upmutators

1. AID/APOBEC cDNA cloned into bacterial expression vector pBAD30 (carries ampicillin resistance gene, *araC*, and the pBAD promoter to control the expression of AID/APOBEC).

2. 20% L-Arabinose (Sigma-Aldrich; w/v) in sterile water stored at –20°C.

3. All materials from Subheading 2.1.

2.5. Lac Reversion Assay

1. Competent CC102.

2. pTrc99a-AID/APOBEC upmutants, wild-type counterparts, and empty vector.

3. LB media.

4. 100 mg/ml ampicillin.

5. 1 M IPTG.

6. M9 + 0.2% lactose agar supplemented with 100 µg/ml ampicillin and 1 mM IPTG.

7. LB agar supplemented with 100 µg/ml ampicillin.

8. 96 deep well, 2 ml capacity plates (Corning).

9. ColiRollers plating beads (Novagen).

2.6. Rifampicin Reversion Assay

Materials from Subheading 2.4 with the following replacements:

1. Competent KL16 (Hfr PO-45 *relA*1 *spoT thi*-1) instead of CC102.

2. LB low salt agar supplemented with 100 µg/ml ampicillin and 50 µg/ml rifampicin instead of M9 + 0.2% lactose agar.

3. Methods

The papillation assay is a widely used technique for the isolation of bacterial colonies exhibiting a high mutation rate (16–21). The principle of the papillation assay relies on using lactose indicator medium such as MacConkey-lactose plates (contains peptone, lactose, bile salts, crystal violet, and neutral red pH indicator), where Lac⁻ bacteria grow as white colonies. Once the peptones are exhausted, the Lac⁻ colonies will stop growing. Only Lac⁺ mutant cells arising within the colony can utilize the lactose in the medium and grow out from the main Lac⁻ colony to form micro-colonies known as papillae. The metabolism of lactose to acidic products decreases the local pH in the media, causing the colour of the neutral red indicator to change to red. In addition, there is surrounding precipitation of protonated bile salts which absorb the neutral red indicator. Thus the Lac⁺ revertants form visually discernable red papillae from the white Lac⁻ colony. Since each Lac⁺ papilla arose from a single mutation event, the number of papillae provides an estimate of the mutation frequency in the colony. A colony with a high rate of mutation will produce many more papillae than a colony with a low rate of mutation.

To assay for transition mutations at C:G basepairs resulting from DNA deamination using the papillation assay, the Lac⁻ strain CC102 was used, which carries a F′ episome encoding *lacZ* containing a mis-sense mutation at residue Glu461 (GAG) to give Gly461 (GGG) (Fig. 1) (22). Reversion back to Lac⁺ is possible via a C to T transition mutation on the corresponding non-coding strand of codon 461 to restore the wild-type glutamate residue.

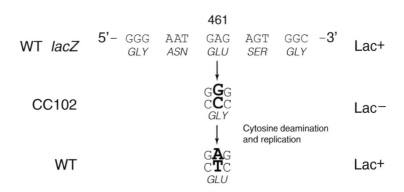

Fig. 1. Lac⁺ reversion of strain CC102 by DNA deamination. The WT *lacZ* region containing codon 461 with the equivalent amino acids are shown. Both coding and non-coding strands of the mutated codon 461 (mutated basepair highlighted in bold) in strain CC102 are below the WT counterpart. DNA deamination of the cytosine at codon 461 on the non-coding strand results in a transition mutation at this C:G basepair such that the codon now encodes glutamate, causing reversion from Lac⁻ back to Lac⁺.

Thus when AID/APOBECs are expressed in CC102, the increased frequency of C to T transition mutations should also be accompanied with increased papillation over background.

3.1. Preparation of MacConkey-Lactose Agar

1. 500 g of MacConkey-lactose dehydrated media is mixed with 1 L of sterile water in a 2-L sterile conical flask.

2. The mixture is brought to boil on a heated plate, and gently boiled for 15 min (see Note 1).

3. The MacConkey-lactose agar is cooled to 50°C in a water bath.

4. 1 ml of 100 mg/ml ampicillin and 1 ml of 1 M IPTG are added to the 1 L MacConkey agar (final concentration 100 μg/ml ampicillin and 1 mM IPTG).

5. Working quickly, the MacConkey agar is poured into sterile dishes. 1 L of MacConkey agar will be enough for ten 145-mm diameter round petri dishes or four 245 mm × 245 mm × 25 mm large square bioassay dishes (see Note 2).

6. The plates are left to solidify at room temperature overnight.

7. The 145-mm round plates are ready to use the following day, but the large square plates will require further incubation at 37°C for up to 48 h to ensure the surface of the agar is dry before use (see Note 3).

8. The large square plates are preferentially used when screening cDNA libraries due to their larger surface area compared to the round plates, thus permitting a greater number of colonies that can be plated.

9. Plates can be stored in the dark at 4°C for up to 2 weeks.

3.2. Papillation Assay to Detect AID/APOBECs

1. *E. coli* strain CC102 is maintained on minimal glucose agar, to select for the F′ episome which contains the genes necessary for proline prototrophy. Thus the loss of the F′ episome will prevent growth on minimal glucose media (lacks proline).

2. The CC102 is made competent for transformation using the Inoue method (23). Other methods of preparing competent *E. coli*, e.g. electro-competence are also acceptable.

3. An aliquot (50 μl) of chemically competent CC102 is mixed with 50 ng of AID/APOBEC cDNA cloned in plasmid pTrc99a in a 1.5-ml Eppendorf tube.

4. The bacteria is then incubated on ice for 30 min before heat shocking at 42°C for 45 s in a water bath before returning to ice for an additional 1 min.

5. To the heat-shocked bacteria, 1 ml of 2× TY media is added and the culture incubated at 37°C for 45 min.

6. The transformed CC102 is diluted appropriately in M9 solution and plated on MacConkey-lactose agar supplemented with 100 µg/ml ampicillin and 1 mM IPTG. Roughly 50–100 colonies are plated onto 145-mm diameter round petri dishes, and 250–500 colonies onto 245 mm × 245 mm × 25 mm large square bioassay dishes (see Note 4).

7. The plates are incubated at 37°C for 3–6 days with papillae becoming visible after 3 days and their numbers increasing up to day 7 (see Note 5 and Fig. 2a).

8. The enhanced papillation produced by colonies can be retested by picking the Lac⁻ portion of the colony and restreaking on fresh MacConkey plates.

9. Papillation can be semi-quantified by counting the average number of papillae per colony from 12 randomly selected colonies after restreaking (Fig. 2b).

10. The identity of the mutation conferring Lac⁺ can be determined by sequencing the relevant section of *lacZ* encompassing codon 461 (primer sequences shown in Subheading 2.2).

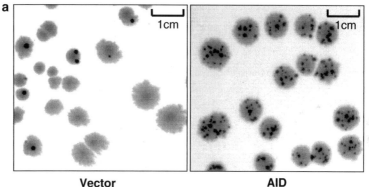

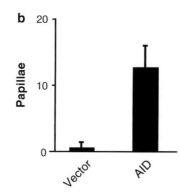

Fig. 2. Assaying AID activity by papillation. (**a**) Empty vector or AID-transformed CC102 were plated on MacConkey-lactose agar and grown at 37°C for 6 days. (**b**) The average number of papillae per colony on MacConkey-lactose agar in CC102 expressing vector control (Vector) or AID (histogram showing the mean and s.d. from 12 randomly selected colonies).

**3.3. Selection for AID/
APOBEC Upmutants**

The papillation assay offers a quick method of screening for variants of AID/APOBECs with enhanced activity from a library of mutants. We have used this technique to isolate AID upmutants with up to 400-fold increased mutator activity above wild-type AID as judged by the frequency of rifampicin resistance (24). There are numerous methods of generating a mutant cDNA library of AID/APOBECs. In the past, we have carried out error-prone PCR mutagenesis using 10 mM MgCl$_2$ and Taq polymerase to generate a random point mutant library of AID (24). The following method describes the generation of such a library and how it can be screened using the papillation assay to isolate upmutants.

1. The following error-prone PCR is prepared in 50 μl volume: 1 ng AID cDNA, 1× Taq buffer, 250 nM of each dNTPs, 10 mM MgCl$_2$, 1 μM of forward and reverse primer, 2.5 U Taq polymerase (Bioline), and made up to 50 μl using sterile water. Primer sequences are shown in Subheading 2.3.

2. PCR is carried out using the following cycling parameters: 94°C, 2 min; then 94°C, 30 s; 65°C, 30 s; 72°C, 1 min for 30 cycles; followed by 72°C, 5 min.

3. The PCR product is cleaned up using the QIAquick PCR Purification Kit (Qiagen), digested with *Eco*RI and *Hin*dIII restriction enzymes (NEB), further purified using the Gel Extraction Kit (Qiagen), and ligated into pTrc99a (pre-digested with *Eco*RI and *Hin*dIII) using Rapid Ligation Kit (Roche). All the reactions are carried out according to the manufacturer's instructions.

4. 10 μl of the ligation reaction mix is transformed into 200 μl of chemically competent DH5α by heat shock as described in Subheading 3.2 and cultured in 1 ml 2× TY at 37°C.

5. The transformation is repeated three times further and all four transformation cultures are pooled to form a single library consisting of 4 ml of culture.

6. 200 μl from the 4 ml culture is used to determine the library titre (see Note 6), and the remainder used to inoculate 250 ml of 2× TY media supplemented with 100 μg/ml ampicillin and cultured overnight at 37°C, followed by plasmid extraction using HiSpeed Maxi Kit (Qiagen) as per manufacturer's instructions.

7. The diversity of the library can be assessed by transforming 1 μl of the library into 100 μl DH5α, plating on LB agar supplemented with 100 μg/ml ampicillin, and sequencing 24–48 randomly selected colonies.

8. Three aliquots (50 μl) of chemically competent CC102 are transformed using pTrc99a containing empty vector (negative control), wild-type AID, and AID mutant cDNA library. The transformation is carried out using heat shock as described in Subheading 3.2.

9. Between 250 and 500 transformed colonies are plated on MacConkey-lactose agar supplemented with 100 μg/ml ampicillin and 1 mM IPTG in 245 mm×245 mm×25 mm large square bioassay dishes and incubated at 37°C until papillation becomes visible.

10. Colonies which give enhanced papillation over the wild-type AID are retested by restreaking on fresh MacConkey plates.

11. To extract the plasmid encoding the AID upmutant, the papillating colony is used to inoculate a 1.5-ml culture of 2× TY media supplemented with 100 μg/ml ampicillin, incubated overnight at 37°C, and the plasmid purified using QIAprep Spin Miniprep Kit (Qiagen) according to the manufacturer's instructions.

12. The identities of the mutations in the AID upmutants are determined by sequencing.

3.4. Using Arabinose Induction to Select for AID/APOBEC Upmutators

From experience, the selection for increasingly more active AID upmutants can be limited by the toxicity of the upmutant when expressed in *E. coli*, and the difficulty in differentiating visually the additional papillation. We have overcome this problem by lowering the expression level of the protein by expressing AID upmutants from the plasmid pBAD30, an arabinose inducible vector (25). This is described in the following method:

1. The AID upmutants are cloned into the plasmid pBAD30 and transformed into CC102 (see Note 7).

2. The transformed CC102 is plated on MacConkey-lactose agar supplemented with L-(+)-arabinose (instead of IPTG) at concentrations ranging from 0.002 to 0.2%.

3. The plates are incubated at 37°C until papillation becomes visible.

4. The level of papillation produced by CC102 colonies is dependent on the amount of arabinose in the media (Fig. 3; see Notes 8 and 9). The concentration of arabinose that results in the formation of three to eight papillae in colonies expressing the AID upmutants can be used in future experiments to induce protein expression in additional rounds of selection for AID upmutants.

3.5. Lac Reversion Assay

To more accurately quantify the mutator frequency of any selected AID upmutant, we use the lac reversion assay and the rifampicin assay. The lac reversion assay is described below:

1. CC102 is transformed with pTrc99a empty vector, or encoding wild-type AID and AID upmutants.

2. Twelve colonies of CC102 transformed with each construct are picked and diluted in 200 μl M9 solution.

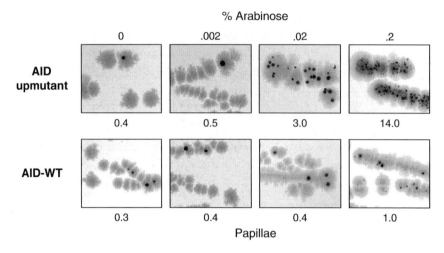

Fig. 3. Papillation by arabinose induction. AID cDNA was cloned into plasmid pBAD30. Expression is driven by the pBAD promoter, which is controlled by the araC molecule encoded on the plasmid (25). Papillation by WT AID and AID upmutant expressed from plasmid pBAD30 as a function of the concentration (% w/v) of L-arabinose (inducer). The average numbers of papillae per colony are shown below.

3. 2 μl of the diluted culture is then used to inoculate 1.5 ml of LB supplemented with 100 μg/ml ampicillin and 1 mM IPTG, and cultured overnight at 37°C.

4. From each of the incubated culture, 100 μl is plated on M9 + 0.2% lactose agar supplemented with 100 μg/ml ampicillin and 1 mM IPTG, and 100 μl of culture diluted to 10^{-6} in M9 solution is plated on LB supplemented with ampicillin to determine culture viability.

5. Plates are incubated between 24 and 48 h at 37°C until colonies become visible.

6. The resulting number of colonies is counted. The frequency of reversion to Lac⁺ is measured by determining the median number of colony-forming cells that grew on lactose selection per 10^{7} viable cells plated with each median determined from 12 independent cultures (see Notes 10 and 11).

7. When testing large number of transformants, 96 × 1.5-ml capacity plates can be used for culturing and ColiRollers plating beads (Novagen) used to speed up plating.

3.6. Rifampicin Assay   This method is similar to the lac reversion assay but quantifies the mutator frequency at a different locus, in this case the *rpoB* gene which when mutated can give rise to resistance to rifampicin. This provides an indication as to whether the AID upmutant is able to globally enhance the rate of mutation in *E. coli*, or preferentially targets the F′ *lacZ* locus. We prefer to perform the rifampicin assay in the *E. coli* strain KL16 rather than CC102 to assess whether differences in the strain could account for the increased activity of the AID upmutants.

1. The first three steps of this method are identical to the lac reversion assay described in Subheading 3.5, but with KL16 used instead of CC102 to be transformed with pTrc99a plasmids.

2. After overnight incubation of the cultures, 300 μl of each culture is plated on LB low salt agar supplemented with 50 μg/ml rifampicin and 100 μg/ml ampicillin and 100 μl of culture diluted to 10^{-6} is plated on LB supplemented with ampicillin to determine culture viability.

3. The remainder of the assay is identical to the lac reversion assay. The frequency of resistance to rifampicin is measured by determining the median number of colony-forming cells that grew on rifampicin selection per 10^7 viable cells plated with each median determined from 12 independent cultures (see Note 12).

4. Notes

1. The MacConkey agar can also be sterilized by autoclaving rather than boiling.

2. Sufficient volume of the MacConkey-lactose media must be poured into the plates to prevent excessive drying and fragmentation of the agar over the 4–6 days of incubation period. Aim for at least 100 ml in the round plates and 250 ml in the large square plates.

3. The MacConkey-lactose agar provides the best results when the surface is dry, thus we often pre-incubate the plates for up to 48 h at 37°C before use. To encourage the drying process, we sometimes remove the lid of the plates and incubate for 6–12 h at 37°C.

4. In our experience, papillation is best visualized on MacConkey-lactose agar when the bacterial colony is able to grow to at least 5 mm in diameter. Thus when performing the experiment for the first time, or using newly prepared competent CC102, we advise plating a range of colony numbers to assess the optimal density for papillation. We typically plate transformed CC102 at serial dilutions ranging from 10^{-1} to 10^{-4}.

5. The number of papillae per colony is the important indicator of mutator frequency. The size of the papillae is irrelevant as each papilla originates from a single bacterium that reverted to Lac⁺, and a large papilla only indicates an early reversion.

6. To estimate the titre of the mutant AID libraries, an aliquot of DH5α culture transformed with the ligation reaction mixture is plated in duplicates at dilution of 10^{-1}, 10^{-2}, and 10^{-3} on LB agar supplemented with 100 μg/ml ampicillin. Plates are incubated at 37°C overnight and the number of colony forming cells is counted. Colony forming units (cfu) per ml is calculated by: titre (cfu per ml) = (cfu × dilution factor)/ volume of culture plated. Total cfu = average titre × total culture volume of library.

7. Both pTrc99a and pBAD30 use *Eco*RI and *Hind*III restriction sites which permits the transfer of cloned cDNAs between the two vectors by digestion and ligation.

8. When using arabinose to induce protein expression in the papillation assay, we found that the average number of papillae per colony for an AID upmutant (K10E, K34E, R157G) could be decreased from 14 to 0.5 by changing the level of induction from 0.2% down to 0.002% arabinose, while wild-type AID induced barely detectable papillation over a similar range of arabinose concentrations (Fig. 3).

9. Although CC102 is Ara⁻, we found that transformation with plasmid pBAD30 (encodes *araC*) could revert the strain to Ara⁺. This becomes significant when raising the arabinose concentration to above 0.2% in MacConkey agar as the colonies now appear entirely red due to the increased metabolism of arabinose, which lowers the pH of the media in similar fashion to lactose metabolism. This makes differentiating visually any Lac⁺ papillae impossible, and we thus recommend not exceeding 0.2% arabinose to induce protein expression.

10. Different sized colonies will appear on the selection plates when performing the lac reversion and the rifampicin assay. We try to count all the colonies irrespective of size.

11. Some cultures will produce an unexpectedly high number of Lac⁺ or rifampicin-resistant colonies. This is mostly likely caused by the clonal expansion of a mutant which arose early during the culturing process (jackpot effect), and does not reflect an intrinsic high mutator frequency. Data collected from these cultures should be excluded from calculating the mean lac reversion or rifampicin-resistant frequencies. To minimize the occurrence and bias of the "jackpot" effect, we try to perform fluctuation analysis using 12 independent cultures, and use a smaller bacterial load to inoculate the cultures by initially diluting the colony in M9 salt solution.

12. We have observed cultures of KL16 or CC102 expressing very active AID upmutants to produce very low viability when plated on LB ampicillin agar following overnight incubation. This can introduce large fluctuations when calculating the

frequency of reversion to lac or resistance to rifampicin. To overcome this problem, we suggest plating at a lower dilution (10^{-4} to 10^{-5} rather than 10^{-6}) onto LB ampicillin plates.

Acknowledgments

We are indebted to J. Miller (Molecular Biology Institute and Department of Biology, University of California, Los Angeles) for kindly providing *E. coli* strain CC102 and recommendations regarding plating, and Gareth Williams and Salome Adam (Cambridge, UK) for helpful comments on the manuscript, and the James Baird and the Frank Elmore funds for support to M.W.

References

1. Neuberger, M. S., Harris, R. S., Di Noia, J., and Petersen-Mahrt, S. K. (2003) Immunity through DNA deamination, *Trends Biochem Sci 28*, 305–312.

2. Malim, M. H. (2009) APOBEC proteins and intrinsic resistance to HIV-1 infection, *Philos Trans R Soc Lond B Biol Sci 364*, 675–687.

3. Navaratnam, N., Morrison, J. R., Bhattacharya, S., Patel, D., Funahashi, T., Giannoni, F., Teng, B. B., Davidson, N. O., and Scott, J. (1993) The p27 catalytic subunit of the apolipoprotein B mRNA editing enzyme is a cytidine deaminase, *J Biol Chem 268*, 20709–20712.

4. Teng, B., Burant, C. F., and Davidson, N. O. (1993) Molecular cloning of an apolipoprotein B messenger RNA editing protein, *Science 260*, 1816–1819.

5. Conticello, S. G., Thomas, C. J., Petersen-Mahrt, S. K., and Neuberger, M. S. (2005) Evolution of the AID/APOBEC family of polynucleotide (deoxy)cytidine deaminases, *Mol Biol Evol 22*, 367–377.

6. Harris, R. S., Bishop, K. N., Sheehy, A. M., Craig, H. M., Petersen-Mahrt, S. K., Watt, I. N., Neuberger, M. S., and Malim, M. H. (2003) DNA deamination mediates innate immunity to retroviral infection, *Cell 113*, 03–809.

7. Coker, H. A., Morgan, H. D., and Petersen-Mahrt, S. K. (2006) Genetic and *in vitro* assays of DNA deamination, *Methods Enzymol 408*, 156–170.

8. Pham, P., Bransteitter, R., Petruska, J., and Goodman, M. F. (2003) Processive AID-catalysed cytosine deamination on single-stranded DNA simulates somatic hypermutation, *Nature 424*, 103–107.

9. Bransteitter, R., Pham, P., Calabrese, P., and Goodman, M. F. (2004) Biochemical analysis of hypermutational targeting by wild type and mutant activation-induced cytidine deaminase, *J Biol Chem 279*, 51612–51621.

10. Shen, H. M. and Storb, U. (2004) Activation-induced cytidine deaminase (AID) can target both DNA strands when the DNA is supercoiled, *Proc Natl Acad Sci USA 101*, 12997–13002.

11. Sohail, A., Klapacz, J., Samaranayake, M., Ullah, A., and Bhagwat, A. S. (2003) Human activation-induced cytidine deaminase causes transcription-dependent, strand-biased C to U deaminations, *Nucleic Acids Res 31*, 2990–2994.

12. Harris, R. S., Petersen-Mahrt, S. K., and Neuberger, M. S. (2002) RNA editing enzyme APOBEC1 and some of its homologs can act as DNA mutators, *Mol Cell 10*, 1247–1253.

13. Petersen-Mahrt, S. K., Harris, R. S., and Neuberger, M. S. (2002) AID mutates *E. coli* suggesting a DNA deamination mechanism for antibody diversification, *Nature 418*, 99–103.

14. Beale, R. C., Petersen-Mahrt, S. K., Watt, I. N., Harris, R. S., Rada, C., and Neuberger, M. S. (2004) Comparison of the differential context-dependence of DNA deamination by APOBEC enzymes: correlation with mutation spectra in vivo, *J Mol Biol 337*, 585–596.

15. Ramiro, A. R., Stavropoulos, P., Jankovic, M., and Nussenzweig, M. C. (2003) Transcription enhances AID-mediated cytidine deamination by exposing single-stranded DNA on the nontemplate strand, *Nat Immunol 4*, 452–456.

16. Konrad, E. B. (1978) Isolation of an *Escherichia coli* K-12 dnaE mutation as a mutator, *J Bacteriol 133*, 1197–1202.

17. Nghiem, Y., Cabrera, M., Cupples, C. G., and Miller, J. H. (1988) The mutY gene: a mutator locus in *Escherichia coli* that generates G.C – T.A transversions, *Proc Natl Acad Sci USA 85*, 2709–2713.

18. Ruiz, S. M., Letourneau, S., and Cupples, C. G. (1993) Isolation and characterization of an *Escherichia coli* strain with a high frequency of C-to-T mutations at 5-methylcytosines, *J Bacteriol 175*, 4985–4989.

19. Yang, H., Wolff, E., Kim, M., Diep, A., and Miller, J. H. (2004) Identification of mutator genes and mutational pathways in *Escherichia coli* using a multicopy cloning approach, *Mol Microbiol 53*, 283–295.

20. Gabrovsky, V., Yamamoto, M. L., and Miller, J. H. (2005) Mutator effects in *Escherichia coli* caused by the expression of specific foreign genes, *J Bacteriol 187*, 5044–5048.

21. Yang, H., To, K. H., Aguila, S. J., and Miller, J. H. (2006) Metagenomic DNA fragments that affect *Escherichia coli* mutational pathways, *Mol Microbiol 61*, 960–977.

22. Cupples, C. G. and Miller, J. H. (1989) A set of lacZ mutations in *Escherichia coli* that allow rapid detection of each of the six base substitutions, *Proc Natl Acad Sci USA 86*, 5345–5349.

23. Sambrook, J. and Russell, D. (2001) *Molecular cloning: a laboratory manual.* Cold Spring Harbor Laboratory Press, Woodbury, NY. 66, 75.

24. Wang, M., Yang, Z., Rada, C., and Neuberger, M. S. (2009) AID upmutants isolated using a high-throughput screen highlight the immunity/cancer balance limiting DNA deaminase activity, *Nat Struct Mol Biol 16*, 769–776.

25. Guzman, L. M., Belin, D., Carson, M. J., and Beckwith, J. (1995) Tight regulation, modulation, and high-level expression by vectors containing the arabinose PBAD promoter, *J Bacteriol 177*, 4121–4130.

Chapter 12

Biochemical Fractionation and Purification of High-Molecular-Mass APOBEC3G Complexes

Ya-Lin Chiu

Abstract

Human APOBEC3G (A3G) is a cytidine deaminase that broadly restricts the replication of many retroviruses, including HIV-1. In different cell types, cytoplasmic A3G is expressed in high-molecular-mass (HMM) RNA–protein complexes or low-molecular-mass (LMM) forms displaying different biological activities. LMM A3G has been proposed to restrict HIV-1 infection soon after virion entry in resting CD4 T cells, monocytes, and mature dendritic cells. Cellular activation and specific cytokine signaling promote the recruitment of LMM A3G into HMM complexes that are likely nucleated by the induced expression of Alu retroelement RNAs. HMM A3G sequesters these retroelement RNAs away from the nuclear LINE-derived enzymes required for Alu retrotransposition. However, assembly of A3G into HMM complexes suppresses its enzymatic activity and may render cells permissive to HIV-1 infection. During HIV-1 virion formation, newly synthesized LMM A3G is preferentially encapsidated when the HIV-1 viral protein viral infectivity factor is absent and employs sequential actions to restrict HIV-1. A3G's biological activities are tightly regulated by its ability to assemble into HMM complexes. Here, we describe in detail the procedures for biochemical fractionation and purification of HMM A3G complexes. Purified HMM A3G complexes will be useful for studying many aspects of the A3G biology, including A3G's roles in restricting retroviral replication, inhibiting retroelement mobility, and potentially regulating cellular RNA function.

Key words: APOBEC3G, Cytidine deaminases, HIV-1, Vif, Alu, Retrotransposition, FPLC, Tandem affinity purification, High-molecular-mass, Low-molecular-mass

1. Introduction

Human cells express a family of enzymes that provide intrinsic immunity against HIV-1. The discovery of these enzymes, APOBEC3G (apolipoprotein B mRNA-editing enzyme, catalytic polypeptide-like 3G, or A3G) (1) and related deoxycytidine deaminases, has fueled intense interest among HIV biologists.

Ruslan Aphasizhev (ed.), *RNA and DNA Editing: Methods and Protocols*, Methods in Molecular Biology, vol. 718,
DOI 10.1007/978-1-61779-018-8_12, © Springer Science+Business Media, LLC 2011

Rapid progress has been made in understanding the antiviral effects of A3G that involve both intravirion and intracellular actions (2).

In the absence of viral infectivity factor (Vif), newly synthesized A3G in virus-producing cells is incorporated into budding virions and employs sequential actions to restrict HIV-1 (3). Initially, A3G bound to HIV-1 RNA may physically impede reverse transcriptase movement on the viral RNA template, resulting in a block in reverse transcription (4, 5). However, this inhibition is frequently incomplete, and minus-strand viral cDNA is generated. Subsequently, A3G deaminase activity induces extensive dC–dU mutations in newly synthesized viral cDNAs (6–9). This action of A3G effectively halts HIV-1 replication because of the resulting dG-to-dA hypermutations in the viral plus-strand DNA, which would contain multiple stop codons or encode altered proteins, or because the uracil-containing minus-strand is destroyed by the actions of uracil DNA glycosylase and apurinic–apyrimidinic endonuclease. Additionally, diminished chromosomal integration of the proviral DNA may occur (10, 11).

HIV-1 utilizes its Vif protein to circumvent the antiviral action of A3G. Vif directly interacts with A3G and links A3G to an E3 ubiquitin ligase complex that mediates polyubiquitylation and accelerated degradation of both A3G and Vif by the 26S proteasome (12–17). Vif also partially impairs the translation of A3G mRNA (12, 17). The combined effects of accelerated degradation and diminished synthesis result in the nearly complete depletion of intracellular A3G in the virus-producing cell and prevent virion encapsidation of A3G, thereby ensuring high infectivity of the progeny virions.

As noted earlier, the antiviral effects of A3G involve both intravirion and intracellular actions. Endogenous cytoplasmic A3G is expressed in two forms in two populations of CD4 T cells and can be resolved by fast protein liquid chromatography (FPLC). Circulating resting CD4 T cells have low-molecular-mass (LMM) A3G and are refractory to HIV-1 infection, at least in part because of an early postentry block (18). In sharp contrast, in some lymphoid tissue-resident resting CD4 T cells, which display increased permissiveness to HIV-1 infection, A3G is predominantly found in high-molecular-mass (HMM) complexes (19). Although the expression of LMM A3G correlates with resistance to HIV-1 infection, the role of LMM A3G as a postentry restriction factor is under debate. An earlier study demonstrated that RNA interference-mediated depletion of LMM A3G in resting CD4 T cells renders these cells permissive for HIV-1 infection (18). It suggested that LMM A3G functions as a potent postentry restriction factor that inhibits the replication of incoming HIV-1 virions in target cells independently of its prior incorporation into virions. Contemporaneous studies also supported the finding and reported that causing LMM A3G to assemble as HMM complexes in naïve resting CD4+ T cells renders the cell more susceptible to HIV infection (19) and that HIV preferentially infects CD4+ T cells containing HMM (rather than LMM) A3G (20).

The postentry antiviral activity of LMM A3G may also govern the resistance of cells of monocyte lineage to HIV-1 infection. Specifically, freshly isolated monocytes are refractory to infection and have LMM A3G (18, 21). Maturation of dendritic cells and interferon treatment of plasmacytoid dendritic cells are associated with additional expression of LMM A3G and less permissiveness to HIV-1 infection (20, 22, 23). Depletion of LMM A3G in dendritic cells also renders these cells permissive for HIV-1 infection (22, 23). Although two studies challenged the role of LMM A3G as a postentry restriction factor in resting CD4 T cells (24, 25) and the aforementioned RNA interference study was subsequently retracted based on inability to reproduce the key findings (18), A3G's postentry restricting activities have been reported in interferon-treated CD4 T cells and activated T helper subtype cells, in which relatively greater amount of LMM A3G proteins are present (26, 27). Another study further demonstrated that CCR6 ligand-mediated induction of A3G expression in activated CD4+ T cells restricts HIV-1 infection at early postentry steps and this restriction was countered by knocking-down the A3G in these cells using RNA interference (28). While the complex status of the CCR6 ligand-induced A3G remained to be addressed, this study supported the finding of A3G's postentry antiviral activity in CD4+ T cells and continued the viability of this topic (28).

Many studies characterizing the HMM A3G complexes by tandem affinity purification (TAP) and mass spectrometry have identified numerous cellular RNA binding proteins with diverse roles in RNA function, metabolism, and fate determination (29–31). Careful analysis suggested that these protein cofactors composed of a complex array of cytoplasmic ribonucleoproteins, including polysome-associated Staufen-containing RNA granules (29), Ro ribonucleoproteins (29–31), and some other RNA-rich cytoplasmic microdomains, such as stress granules (30, 31) and processing bodies (30, 32). Potential functions of A3G in counteracting the inhibition of protein synthesis by various microRNAs and in facilitating the association of microRNA-targeted mRNA with polysomes have also been reported (33).

Intriguingly, sequencing of the most prominent RNA components in HMM A3G complexes identified human endogenous Alu retroelement RNAs and small cytoplasmic hY RNAs and suggested that endogenous nonautonomous retroelements (i.e., Alu RNAs) are the natural cellular targets of A3G (29). Alu RNAs are the most prominent nonautonomous mobile genetic elements in human cells. They amplify and mobilize through retrotransposition, an intracellular process that involves cytoplasmic RNA intermediates, reverse transcription in the nucleus, and integration of the newly formed retroelement DNA at novel chromosomal sites. Alu elements encode no protein, and their retrotransposition depends on the nuclear enzymatic machinery encoded by a set of autonomous retroelements termed long interspersed nuclear elements-1 (LINE-1, L1) (34, 35).

When tested in an *in vitro* assay measuring Alu retrotransposition in living cells, A3G greatly inhibited L1-dependent retrotransposition of Alu elements without affecting the retrotransposition of L1 (29, 36). Since A3G is primarily cytoplasmic and Alu RNA is recruited to Staufen RNA granules in an A3G-dependent manner, A3G likely interrupts retrotransposition by sequestering transcribed Alu RNAs in the cytoplasm, denying Alu RNAs access to the nuclear enzymatic machinery of L1 (29). Unfortunately, the assembly of HMM A3G complexes to combat Alu retrotransposition opens the door for HIV-1 infection, as the postentry restricting activity of LMM A3G is forfeited.

In summary, the recruitment of A3G into LMM and HMM complexes highlights two opposing functional conformations of this protein, as it balances its biological activities against exogenous viruses and endogenous mobile genetic elements for optimum benefit for human cells. To facilitate the study of A3G's roles as a restriction factor of retroviral replication, as an inhibitor of retroelement mobility, and as a potential regulator of cellular RNA function, we describe here in detail the procedures for FPLC fractionation to resolve LMM and HMM A3G complexes and for TAP of HMM A3G complexes.

2. Materials

2.1. Human Primary Cell Isolation

1. Human buffy coat (Stanford School of Medicine Blood Center, Palo Alto, CA).

2. MACS buffer: phosphate-buffered saline (PBS), pH 7.2, supplemented with 2% fetal bovine serum (FBS) (Gemini, West Sacramento, CA) and 2 mM EDTA.

3. RPMI 1640 medium supplemented with 10% FBS, 100 U/ml penicillin, 100 μg/ml streptomycin, and 2 mM L-glutamine.

4. Macrophage medium: RPMI 1640 supplemented with 10% FBS, 10% human AB serum (Gemini), 100 U/ml penicillin, 100 μg/ml streptomycin, and 2 mM L-glutamine.

5. Ficoll-Paque Plus (GE Healthcare, Piscataway, NJ).

6. MACS CD14 and CD4 microbeads (Miltenyi Biotec, Bergisch Gladback, Germany).

7. MACS LS columns (Miltenyi Biotec, Bergisch Gladback, Germany).

8. Human interleukin 2 (IL-2) (Roche, Indianapolis, IN) is stored in single-use aliquots at −30°C.

9. Lectin (Erythroagglutinin PHA-E, Sigma-Aldrich, St. Louis, MO) is dissolved at 1 mg/ml in PBS and stored in single-use aliquots at −30°C.

2.2. Cell Culture and Transfection

1. Dulbecco's modified Eagle's medium (DMEM) (Gibco/BRL, Bethesda, MD) supplemented with FBS, 100 U/ml penicillin, 100 µg/ml streptomycin, and 2 mM L-glutamine.

2. Solution of trypsin (0.05%) and EDTA (0.53 mM) (Gibco/BRL).

3. 2× Hank's balanced salt solution: 280 mM NaCl, 10 mM KCl, 1.5 mM Na_2HPO_4, 12 mM dextrose, 50 mM HEPES, pH 7.05. Sterilize by filtration through a 0.22-µm membrane.

4. 2 M $CaCl_2$ solution. Sterilize by filtration through a 0.22-µm membrane.

5. 1× Tris/EDTA (TE) buffer: 10 mM Tris–HCl, 1 mM EDTA, pH 8.0. Sterilize the solution by filtration through a 0.22-µm membrane.

6. Appropriate DNA plasmids such as plasmid encoding N-terminal tandem affinity (NTAP)-tagged A3G (NTAP-A3G) or other NTAP-tagged proteins.

2.3. Cell Lysate Preparation and Quantitation

1. 100× EDTA-free protease inhibitor cocktail set V (CalBiochem, Gibbstown, NJ): reconstitute each vial with 1 ml DEPC-treated H_2O to obtain a 100× stock solution. 1× solution contains 500 µM AEBSF-HCl, 150 nM aprotinin, 1 µM E-64, and 1 µM leupeptin hemisulfate.

2. Cell lysis buffer: 50 mM HEPES, pH 7.4, 125 mM NaCl, 0.2% NP-40, store at 4°C. Sterilize the buffer by filtration through a 0.22-µm membrane. Supplement the buffer with 1× EDTA-free protease inhibitor cocktail set V and 0.1 mM phenylmethylsulfonyl fluoride (PMSF) immediately before use. For TAP, increase PMSF to 1 mM.

3. Bio-Rad DC Protein Assay Kit II with bovine serum albumin (BSA) standard.

2.4. Fast-Protein Liquid Chromatography

1. FPLC buffer: 50 mM HEPES, pH 7.4, 125 mM NaCl, 0.1% NP-40, 1 mM dithiothreitol, and 10% glycerol. Sterilize the buffer by filtration through a 0.22-µm membrane twice, degas, and chill to 4°C before use.

2. Superose 6 HR 10/30 gel filtration columns or Superose 6 10/300 GL Tricon high-performance columns (GE Healthcare).

3. AKTA FPLC Chromatographic System (GE Healthcare) in cold room.

4. Gel filtration standards (Bio-Rad, Hercules, CA): Rehydrate the vial of lyophilized mixture of molecular weight markers (range 1,350–670,000 Da) by adding 0.5 ml of DEPC-H$_2$O. Swirl gently to mix and allow the vial to stand in ice for 2–3 min. Swirl the vial again and then centrifuge at $14,000 \times g$ for 5 min in a microfuge at 4°C to remove any fine particulates. Store the standards in single-use aliquots (125 µl) at –80°C.

2.5. Tandem Affinity Purification

1. InterPlay TAP purification kit (Agilent Technologies-Stratagene, La Jolla, CA).

2. Microcon YM-3 centrifugal filter unit (Millipore).

2.6. SDS-Polyacrylamide Gel Electrophoresis

1. Criterion electrophoresis cell (Bio-Rad).

2. 26-Well 12.5% Tris–HCl precast polyacrylamide gels (Bio-Rad).

3. 18-Well 10% Tris–HCl precast polyacrylamide gels (Bio-Rad).

4. 4× SDS sample buffer: 250 mM Tris–HCl, pH 6.8, 8% w/v SDS, 40% (v/v) glycerol, 0.04% (w/v) bromophenol blue, 20% (v/v) β-mercaptoethanol (β-ME). Only add β-ME immediately before use.

5. Running buffer: 25 mM Tris, pH 8.3, 192 mM glycine, 0.1% SDS.

6. SeeBlue Plus2 pre-stained standard (Invitrogen, Carlsbad, CA).

2.7. Silver Staining and Coomassie Blue Staining

1. Ethanol.

2. Acetic acid.

3. SilverQuest silver staining kit (Invitrogen).

4. Pierce Imperial Protein Stain solution (Thermo Scientific, Waltham, MA).

2.8. Western Blotting for APOBEC3G and Other Protein Cofactors

1. Criterion blotter with plate electrodes (Bio-Rad).

2. Immobilon-P transfer membrane (PVDF, 0.45 µm, Millipore, Billerica, MA).

3. Transfer buffer: 25 mM Tris, pH 8.3, 192 mM glycine, 20% (v/v) methanol.

4. Wash buffer (PBST): 1× PBS with 0.1% Tween.

5. Blocking buffer: PBST with 5% nonfat milk.

6. Primary antibody dilution buffer: PBST with 2% BSA and 0.02% w/v sodium azide. Sterilize the buffer by 0.22-µm membrane filtration.

7. Primary antibody: anti-A3G (NIH AIDS Research and Reference Reagent Program, catalog No. 9968, 1:2,500 dilution).

8. Secondary antibody dilution buffer: PBST with 2.5% nonfat milk.

9. Secondary antibodies: anti-rabbit IgG horse-radish peroxidase-linked whole antibody (GE Healthcare, formerly Amersham Biosciences).

10. Western Lightning Chemiluminescence Reagent Plus kit (PerkinElmer, Waltham, MA).

11. Hyperfilm ECL 8 × 10 in. (GE Healthcare).

12. Stripping buffer: 62.5 mM Tris–HCl, pH 6.8, 2% (w/v) SDS. Store at room temperature. Warm to working temperature of 70°C and add 100 mM β-ME before use.

3. Methods

3.1. Human Primary Cell Isolation

When working with human buffy coats, peripheral blood mononuclear cells (PBMC) should be isolated by density gradient centrifugation with Ficoll-Paque Plus (37, 38). The CD14 monocytes and CD4 T cells are purified from PBMC sequentially by selection on CD14 microbeads and CD4 microbeads, respectively.

3.1.1. Density Gradient Centrifugation

1. Aliquot one buffy coat (~30–35 ml) into two 50-ml conical tubes.

2. Suspend the cells gently in MACS buffer using a serological pipette and bring up the volume to 35 ml.

3. Underlay the diluted solution of cells with 10 ml of Ficoll-Paque Plus (see Note 1).

4. Centrifuge at $600 \times g$ (Beckman Coulter Allegra 6R or equivalent) for 30 min at 20°C (see Note 2). No brake.

5. Carefully aspirate the upper layer, leaving the lymphocyte layer undisturbed at the interface.

6. Use a clean pipette to transfer the lymphocyte layer to a clean 50-ml conical tube.

7. To remove platelets after density gradient separation, add MACS buffer to 15 ml and resuspend the cells by gently drawing them in and out with a serological pipette. Then add more MACS buffer to bring up the volume to 50 ml.

8. Centrifuge at $600 \times g$ (Sorvall Legent RT or equivalent) for 5 min at 4°C.

9. Carefully aspirate supernatant. Repeat washing step two more times. Before the final centrifugation, aliquot some cells to determine cell number with hemocytometer.

3.1.2. CD14 Monocyte Isolation

1. Resuspend cell pellet in 900 µl per 10^8 cells and add 100 µl of CD14 microbeads per 10^8 cells. Mix well and incubate for 15 min at 4°C (see Note 3).

2. Wash cells by adding 10 ml of MACS buffer per 10^8 cells and centrifuge at $600 \times g$ for 5 min at 4°C. Aspirate supernatant completely.

3. Resuspend cells in MACS buffer to final concentration of 2×10^8 cells/ml. It is important to obtain a single-cell suspension before magnetic separation.

4. Proceed to magnetic separation with manual LS columns or autoMACS Separator. Follow exactly the instructions provided by Miltenyi Biotec or steps 5–10 below.

5. Place LS column in the magnetic field of a MACS separator compatible with LS column. Rinse the column with 3 ml of MACS buffer.

6. Apply cell suspension to the column.

7. Collect unlabeled cells that pass through the column and wash column three times with 3 ml of MACS buffer each time. Only add new buffer when the column reservoir is empty. Collect total effluent; this is the unlabeled cell fraction (peripheral blood lymphocytes, PBLs) and is the starting material for isolation of CD4 T cells.

8. Remove the column from separator and place it on a 15-ml conical tube.

9. Pipette 5 ml of MACS buffer onto the column and immediately flush out the magnetically labeled cells by firmly pushing the plunger into the column. Determine the cell number with a hemocytometer.

10. Collect the labeled cells (monocytes) by centrifugation at $600 \times g$ for 5 min at 4°C. Aspirate supernatant completely and wash once with PBS. Flash-freeze the cell pellet in liquid nitrogen for later cell lysate preparation. Otherwise, proceed to macrophage differentiation.

11. Macrophages are prepared by culturing monocytes on Primaria-coated 100-mm plates (BD Falcon, Franklin Lakes, NJ) at 10^6 cells/ml in macrophage medium for 10 days.

3.1.3. CD4 T-Cell Isolation

1. CD4 T cells are isolated from human PBLs (unlabeled cells from Subheading 3.1.2, step 7) by incubating with CD4 microbeads. Resuspend the pellet of PBLs in 800 µl per 10^8 cells and add 200 µl of CD14 microbeads per 10^8 cells. Mix well and incubate for 15 min at 4°C (see Note 3).

2. Proceed to wash and magnetic separation with manual LS columns exactly as described in Subheading 3.1.2, steps 2–9.

3. Culture CD4 T cells at 2×10^6 cells/ml in regular supplemented RPMI medium in T75 or T175 flasks depending on the total volume. During the first 2 h of culture, change flasks every hour to completely remove the residual monocytes that co-purify with CD4 T cells and attach to the bottom of the plastic plates (see Note 4). The resulting resting CD4 T cells (>99.5% pure) can be maintained for up to a week before analysis.

4. To activate CD4 T cells, treat them with phytohemagglutinin (5 µg/ml) for 36 h and then with IL-2 (20 U/ml) for 36 h.

3.2. Cell Culture and Transfection

1. H9 human T-lymphoid cells are cultured in regular supplemented RPMI 1640 (30 ml per T75 flask) and are passaged at 1:20 dilution when the medium has turned yellowish.

2. 293T cells are cultured in regular supplemented DMEM and are passaged when approaching confluence with trypsin/EDTA to provide new maintenance cultures in T75 flask.

3. To passage 293T cells, first aspirate the old medium and wash the cells quickly with 1.5 ml of trypsin and immediately remove the trypsin by aspiration. Add another 1.5 ml of trypsin and shake the flask until all the cells are dislodged. Add 8.5 ml of medium to stop the effect of trypsin. Transfer the cell-containing medium to a 15-ml conical tube. Spin down the cells by $600 \times g$ for 5 min at 4°C. Last, remove the supernatant by aspiration and resuspend the cells in 10 ml of new medium. To the new T75 flasks, add 15 ml of regular supplemented DMEM and 0.3 ml of cell suspension for maintenance cultures.

4. To prepare experimental cultures for TAP, collect cells from two T75 flasks (~90% confluence) and seed them into five T175 flasks containing 30 ml of medium per flask. Such passage will provide experimental cultures that approach ~70% confluence and are ready for transfection after 24 h.

5. For each transfection, put 1.5 ml of 0.1× TE buffer in a sterile tube and add appropriate DNAs. For example, 8 µg NTAP-A3G for experimental reaction (see Note 5) and 8 µg of NTAP empty vector and nontagged A3G expression plasmids for control reaction. Add 1.75 ml of 2× Hank's balanced salt solution and mix contents of each tube. Add 220 µl of 2.5 M $CaCl_2$ to each tube and mix each tube immediately.

6. Let the transfection mixture sit for 20 min at room temperature. The solution will gradually become cloudy.

7. Add the transfection mixture to the T175 experimental culture in dropwise fashion, mix by gently rocking back and forth (do not swirl), and culture in 5% CO_2 incubator.

8. The next day (~16 h), replace the medium with 30 ml of fresh regular supplemented DMEM.

9. Cells are typically harvested approximately 48 h after transfection. To harvest the cells, aspirate the medium and add 10 ml of PBS to each flask. Gently shake the flask until all the cells are dislodged.

10. Transfer cell suspension from five T175 flasks (5×10 ml) to a 50-ml conical tube. Centrifuge at $600 \times g$ for 5 min at 4°C. Carefully remove the supernatant by aspiration.

11. Flash-freeze the cell pellet in the tube using liquid nitrogen. The pellet can be stored at -80°C for up to 2 weeks. Cells from five T175 flasks will be the starting material for one reaction of TAP to isolate HMM A3G complexes.

3.3. Cell Lysate Preparation and Quantitation

Extracts are prepared by hypotonically lysing cells and then pelleting and discarding the nuclei and cell debris. Perform lysate preparation steps at 4°C to prevent the HMM complexes from disassembling.

1. For CD4 primary T cells and H9 T cell lines, spin down the cells in 15-ml conical tubes for 5 min at $600 \times g$ at 4°C. Aspirate the supernatant.

2. Resuspend the cells in 1 ml of PBS and transfer the suspension into Eppendorf tubes. Spin down the cells at $1,500 \times g$ (Eppendorf centrifuge 5417R) for 5 min at 4°C. Aspirate the supernatant.

3. Resuspend the cells in ice-cold lysis buffer supplemented with 1× EDTA-free protease inhibitors (see Note 6) and 0.1 mM PMSF (see Note 7). Vortex the suspension for 30 s.

4. For frozen monocytes pellet, add 1 ml of ice-cold lysis buffer supplemented with protease inhibitors. Transfer the suspension into Eppendorf tubes and vortex the suspension for 30 s.

5. For frozen macrophages on 100-mm plates, add 2 ml of ice-cold lysis buffer supplemented with 1× EDTA-free protease inhibitors. Transfer the suspension into Eppendorf tubes and vortex the suspension for 30 s.

6. Rock the suspension using rugged rotator for 30 min in the cold room.

7. Centrifuge the extracts at $14,000 \times g$ (Eppendorf centrifuge 5417R) for 30 min at 4°C.

8. Transfer the supernatant (clarified cytoplasmic lysate) to new Eppendorf tubes. Take a 20-μl aliquot for quantitation with the DC protein assay kit with BSA standard, following the instructions provided by the manufacturer. The lysate is now ready for FPLC analysis. Otherwise, flash-freeze the lysate in liquid nitrogen and store it at -80°C for up to 2 weeks for later analysis.

**3.4. Fast-Protein
Liquid
Chromatography**

Clarified cytoplasmic lysate are applied to a calibrated Superose 6
HR 10/30 gel-filtration column or a Superose 6 10/300 GL
Tricon high-performance columns run by an AKTA FPLC apparatus. In general, to detect endogenous A3G, at least 100–200 µg
of lysate (about 10–20 million CD4 T cells) is needed for each
FPLC run. This condition allows 5–10 different cytokine or stimuli
treatments to the cells isolated from one blood donor. Higher
amounts of lysate (up to 1 mg) can be loaded for each FPLC run
if the fractions will be subjected to further immunoprecipitation
analysis. One column volume (28 ml) is eluted and collected in
1-ml aliquots. A typical example of the results is shown in Fig. 1.
Following is our protocol for size fractionation. Perform fractionation steps at 4°C to prevent the HMM complexes from disas-
sembly (see Note 7).

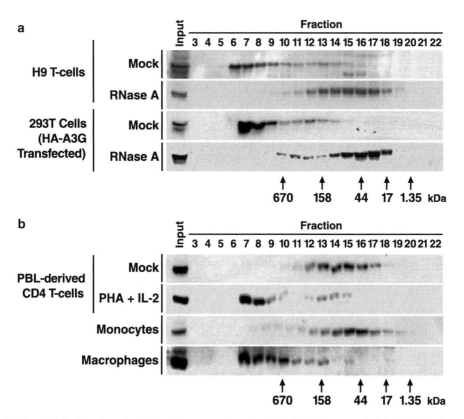

Fig. 1. Biochemical fractionation of cytoplasmic A3G complexes by FPLC. (a) Endogenous A3G (~46-kDa) in human H9
T-cells and exogenous HA-A3G (~50-kDa) expressed in 293T cells principally reside in >700-kDa HMM complexes (fractions 6–8) that are converted to a 46- to 100-kDa LMM forms (fractions 15–17) after treatment with 50 µg/ml RNase A
for 1 h at 4°C. (b) Human peripheral blood-derived resting CD4 T cells and freshly isolated monocytes express LMM A3G.
T-cell activation by PHA and IL-2 treatment and monocyte differentiation into macrophages promote the assembly of
HMM A3G complexes. A3G complexes were resolved by FPLC followed by SDS-PAGE and identified by immunoblotting
with anti-A3G. This figure was adapted in part from ref. (18).

1. Equilibrate sizing column with 2 column volumes (1 column volume = ~28 ml) of FPLC buffer. Operate at a flow rate of 0.15–0.2 ml/min.

2. Load a 1-ml sample loop with 1 ml of clarified cytoplasmic lysate prepared as described in Subheading 3.3. Save another 100 µl as input for later Western analysis. Make sure that the protein concentration of the lysate is within the range of 0.1–1 µg/µl. When the fractionation pattern of A3G complexes in different lysates need to be compared, make sure that all the lysates have been diluted with cell lysis buffer to the same concentration but still within the range of 0.1–1 µg/µl. For column calibration, inject 125 µl of gel filtration standards.

3. Run a program with a 0.15-ml/min flow rate; pressure limit: 1.2 MPa; UV: 10; equilibrate: 1 column volume; empty loop by injecting 1.2 ml of buffer through the sample loop; elution: 1 column volume of buffer; collect elute fraction in 1-ml aliquots.

4. Flash-freeze the elute fractions (1–28) in liquid nitrogen and store them at –80°C for later analysis by SDS-polyacrylamide gel electrophoresis (SDS-PAGE), Western blotting, and immunoprecipitation.

5. Repeat steps 1–4 to analyze multiple samples sequentially.

3.5. Tandem Affinity Purification

Derived from the system described by the Seraphin laboratory, TAP is a two-step affinity purification scheme, which generates protein complexes sufficiently clean to be analyzed by mass spectrometry (39, 40). TAP technology has proven useful in probing interactions in yeast and can be successfully adapted to various organisms including mammalian cells.

Transfection of DNA plasmids encoding NTAP-A3G into 293T cells as described in Subheading 3.2 leads to the expression of a A3G protein fused to two affinity tags, a streptavidin binding peptide and a calmodulin binding peptide. Tagged A3G proteins expressed in 293T cells also form HMM complexes as demonstrated by FPLC analysis (Fig. 1a). TAP of tagged A3G and interacting proteins with the InterPlay TAP purification kit (Stratagene) requires two sequential purification steps that involve gentle washing and small-molecule elution conditions. For the first purification step, apply the cell lysate to a streptavidin resin and then elute with biotin (41, 42). Apply that eluate to a calmodulin resin for the second purification step, and release the protein of interest and its binding partners by removing calcium from the affinity resin with a chelator such as EGTA (43). The pattern of proteins copurifying with NTAP-A3G based on the manual provided by the manufacturer with minor modification proved highly reproducible in multiple independent experiments. A typical example of the results is shown in Fig. 2.

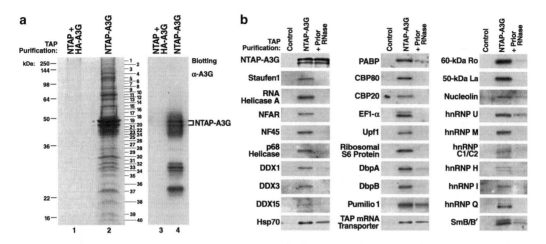

Fig. 2. Purification and characterization of HMM NTAP-A3G complexes. (**a**) HMM NTAP-A3G complexes were purified by TAP, resolved by SDS-PAGE, and visualized by Coomassie staining (*left*) and anti-A3G immunoblotting (*right*). Control cell lysates containing unlinked NTAP and HA-A3G were identically processed (lane 1). The NTAP-A3G gel (lane 2) was cut into 40 slices, digested in-gel with trypsin, and analyzed by tandem mass spectrometry. (**b**) Verifying cofactors identified by tandem mass spectrometry. Purified complexes (NTAP-A3G) or control purifications (unlinked NTAP + HA-A3G) were immunoblotted with antibodies specific for the indicated proteins. To test RNase A sensitivity of cofactor binding in HMM A3G complexes, lysates were pretreated with 50 μg/ml RNase A (+Prior RNase) for 2 h at 4°C before purification. Shown are 30 representative components of 95 previously reported protein cofactors of the HMM A3G complexes. This figure was adapted in part from ref. (29).

3.5.1. Preparing the Buffers for the TAP Protocol

Prepare only the amount of buffers required for 1 day of experimental work and keep them at 4°C. The protease inhibitor cocktail and PMSF must be added to the buffers immediately before use. When processing multiple reactions, reagents may be scaled-up, pooled, and processed concurrently.

1. Cell lysis buffer: Prepare lysis buffer as described in Subheading 2.3 but increase the PMSF to 1 mM.

2. Streptavidin binding buffer (SBB): Prepare the SBB by adding 7 μl of 14.4 M β-mercaptoethanol (β-ME, provided by the kit), 100 μl of 100× protease inhibitor cocktail, and 100 μl of 100 mM of PMSF to 10 ml of SBB solution provided in the kit.

3. Streptavidin elution buffer (SEB): Prepare the SEB by 7 μl of 14.4 M β-ME, 100 μl of 100× protease inhibitor cocktail, and 100 μl of 100 mM PMSF to 10 ml of SEB solution provided in the kit. The SEB must be protected from the light.

4. Calmodulin binding buffer (CBB): Prepare the CBB by adding 7 μl of 14.4 M β-ME, 100 μl of 100× protease inhibitor cocktail, and 100 μl of 100 mM of PMSF to 10 ml of CBB solution provided in the kit.

5. Calmodulin elution buffer (CEB): Prepare the CEB by adding 7 μl of 14.4 M β-ME to 10 ml of CEB solution provided

in the kit. Do not add protease inhibitor to the CEB if the purified proteins are for mass spectrometry analysis.

3.5.2. Preparing the Cell Extracts

1. Add 10 ml of cell lysis buffer to the frozen 293T cell pellet prepared as described in Subheading 3.2. Resuspend the cells and vortex the suspension for 30 s (see Notes 7 and 8).

2. Rock the suspension using a rugged rotator for 30 min in the cold room.

3. Aliquot the extracts into eight Eppendorf tubes. Centrifuge the extracts at $14,000 \times g$ at 4°C for 30 min.

4. Transfer all the supernatant (clarified cytoplasmic lysate) to a new 15-ml conical tube. Save an aliquot of the supernatant for Western analysis (see Note 9). To each 10 ml of cell lysate, add 7 µl of 14.4 M β-ME.

3.5.3. Preparing the Streptavidin Resin

1. Centrifuge 500 µl of the 50% streptavidin resin slurry (provided in the kit) at $1,500 \times g$ for 5 min in a microfuge to collect the resin. Discard the supernatant to remove the ethanol storage buffer.

2. Resuspend the resin in 1 ml of SBB. Collect the resin by centrifugation at $1,500 \times g$ for 5 min using microfuge. Discard the supernatant and repeat this wash step.

3. Collect the resin by centrifugation at $1,500 \times g$ for 5 min in a microfuge. Discard the supernatant and add 250 µl of SBB to the resin resulting in 500 µl of washed streptavidin resin slurry (50%).

3.5.4. Purifying the Protein Complexes Using Streptavidin Resin

1. Add the 500 µl of washed streptavidin resin slurry to the lysate from Subheading 3.5.2, step 4.

2. Rotate the tube at 4°C for 2 h to allow the tagged proteins to bind to the resin.

3. Collect the resin by centrifugation at $300 \times g$ at 4°C for 5 min. Remove the supernatant by careful aspiration. Save an aliquot of the supernatant for post-binding Western analysis (see Note 9).

4. Resuspend the resin in 1 ml of SBB and transfer the suspension to a new Eppendorf tube. Rotate the tube at 4°C for 10 min. Repeat this wash step. Save an aliquot of the beads for pre-elution analysis (see Note 10).

5. Collect the resin by centrifugation at $1,500 \times g$ at 4°C for 5 min. Discard the supernatant. Add 1 ml of SEB to the resin.

6. Rotate the tube at 4°C for 30 min to elute the protein complexes.

7. Collect the resin by centrifugation at $1,500 \times g$ at 4°C for 5 min. Carefully transfer the supernatant to a new 5-ml

polystyrene round-bottom tube. The supernatant (elution 1) contains the eluted proteins. Save an aliquot of the eluate for Western analysis (see Note 9). Save the beads for post-elution analysis (see Note 10).

8. Add 20 µl streptavidin supernatant supplement (provided by the kit) to the supernatant. Then, add 4 ml CBB to the supplemented supernatant.

3.5.5. Preparing the Calmodulin Resin

1. Centrifuge 250 µl of the 50% calmodulin resin slurry (provided in the kit) at $1,500 \times g$ for 5 min in a microfuge to collect the resin. Discard the supernatant to remove the ethanol storage buffer.

2. Resuspend the resin in 1 ml of CBB. Collect the resin by centrifugation at $1,500 \times g$ for 5 min in a microfuge. Discard the supernatant and repeat this wash step.

3. Collect the resin by centrifugation at $1,500 \times g$ for 5 min in a microfuge. Discard the supernatant and add 125 µl CBB to the resin resulting in 250 µl of washed calmodulin resin slurry (50%).

3.5.6. Purifying the Protein Complexes Using Calmodulin Resin

1. Add the 250 µl of washed calmodulin resin slurry to the supplemented supernatant from Subheading 3.5.4, step 8. This supernatant contains the eluted proteins from the streptavidin resin purification.

2. Rotate the tube at 4°C for 2 h to allow the tagged proteins to bind to the resin.

3. Collect the resin by centrifugation at $300 \times g$ at 4°C for 5 min. Remove the supernatant by careful aspiration. Save an aliquot of the supernatant for post-binding analysis (see Note 9).

4. Resuspend the resin in 1 ml of CBB and transfer the suspension to a new Eppendorf tube. Rotate the tube at 4°C for 10 min. Repeat this wash step. Save an aliquot of the beads for pre-elution analysis (see Note 10).

5. Collect the resin by centrifugation at $1,500 \times g$ at 4°C for 5 min. Discard the supernatant. Add 250 µl of CEB to the resin.

6. Rotate the tube at 4°C for 30 min to elute the protein complexes.

7. Collect the resin by centrifugation at $1,500 \times g$ at 4°C for 5 min. Carefully transfer the supernatant to a new Eppendorf tube. The supernatant (elution 2) contains the TAP-purified protein complexes that may be analyzed by Western blotting and silver staining. Save the beads for post-elution analysis (see Note 10). For Coomassie staining and mass spectrometry, the purified proteins need to be concentrated tenfold with a Microcon

YM-3 Centrifugal Filter unit. Consulting the manufacturer's product specifications and guidelines is recommended.

8. Flash-freeze the TAP-purified protein complexes in liquid nitrogen and store it at –80°C for future analysis.

3.6. SDS-Polyacrylamide Gel Electrophoresis

To detect and to characterize the TAP-purified proteins, resolve the protein preparation by SDS-PAGE. These instructions assume the use of Bio-Rad Criterion electrophoresis cells and precast Tris–HCl polyacrylamide gels.

1. To analyze the FPLC fractions by Western blotting, use 26-well 12.5% Tris–HCl polyacrylamide gels. Mix a 15-μl aliquot of the input lysate and collected FPLC fractions (3–26) with 5 μl of 4× sample buffer and then load 15 μl of the 20-μl sample mixture into each well. There is no need to heat the sample mixture before loading.

2. To analyze the TAP-purified protein complexes, use 18-well gels (12.5% gels for Western blotting analysis and 10% gels for silver staining and Coomassie blue staining). For Western blotting and silver staining, mix a 15-μl aliquot of the purified protein complex with 5 μl of 4× sample buffer. For Coomassie blue staining and mass spectrometry, mix ~20 μl of the concentrated protein complexes (described in Subheading 3.5.6, step 7) with 7 μl of 4× sample buffer. Heat these sample mixtures at 90°C for 5 min before loading.

3. Remove the precast Criterion gel cassette from the storage container. Gently remove the comb and rinse the wells thoroughly with running buffer. Remove the tape from the bottom of the cassette. Insert the gel cassette into one of the slots in the Criterion gel and make sure that the upper buffer chamber is facing toward the center of the cell. Repeat for second gel.

4. Fill the upper buffer chamber in each Criterion gel with 60 ml of running buffer. Fill the tank to the line molded into the sides of the tank (~800 ml) or the lower edge of the gels' upper buffer chamber.

5. Load the samples into the wells with a pipette using gel-loading tips. Include one well for prestained molecular weight standards.

6. Put on the tank lid and plug the cables into the power supply with proper polarity. Run the gels at 100 V until pre-stained standards start to resolve then at 150 V until the dye reaches the bottom of the gels.

7. Upon completion of the run, disconnect the gel unit from the power supply. To remove the gel from the cassette, follow exactly the instructions in the Criterion Cell Instruction

Manual provided by the manufacturer. Clean the Criterion tank and lid with deionized water.

8. Proceed to gel staining or Western blotting.

3.7. Silver Staining and Coomassie Blue Staining

To visualize the proteins resolved by SDS-PAGE, silver staining and Coomassie blue staining are performed with the SilverQuest silver staining kit (Invitrogen) and Pierce Imperial Protein Stain solution (Thermo Scientific), respectively. Follow exactly the instruction manuals provided by the manufacturers. Typically, protein bands that are visible by Imperial Protein Stain will result in detectable levels of proteins for tandem mass spectrometry analysis. An example of the Coomassie blue-stained gel is shown in Fig. 2a.

3.8. Western Blotting for APOBEC3G and Other Protein Cofactors

These directions assume the use of the Bio-Rad Criterion Blotter with plate electrodes (tank transfer system) and the Immobilon-P (PVDF) transfer membrane.

1. Prepare a PVDF membrane, two filter papers, and two fiber pads for each gel. The membrane and the filter papers need to be slightly larger than the gel.

2. Wet the PVDF membrane in 100% methanol for 15 s and then transfer it to Milli-Q water for 2 min.

3. Pre-equilibrate PVDF membrane, gel, filter papers, and fiber pads in transfer buffer on a rocking platform for 10 min at room temperature.

4. Prepare gel sandwich in the transfer cassette. The polarity should be cathode–fiber pad–filter paper–gel–PVDF–filter paper–fiber pad–anode.

5. Close the transfer cassette and place it in the tank module and make sure the polarity is in correct direction. Repeat for the other cassette.

6. Put the cooling unit (provided by the manufacturer; needs to be pre-cooled at −30°C) in the tank and completely fill the tank with transfer buffer.

7. Add a standard stir bar in the tank and set the speed as fast as possible to help maintain even buffer temperature and ion distribution in the tank.

8. Put on the transfer tank lid, plug the cables into the power supply, and run the blot at 50 V for 3 h or 30 V overnight in the cold room.

9. Upon completion of the run, disassemble the sandwich and remove the membrane for development. The colored prestained standards should be clearly visible on the membrane. Clean the transfer tank, cassettes and fiber pads, and rinse well with deionized water.

10. Blocking: incubate the membrane in 50 ml of blocking buffer for 1 h at room temperature on a rocking platform. Then, quickly rinse the membrane with 50 ml of washing buffer for 5 min at room temperature on a rocking platform. Repeat this rinse step two more times.

11. Primary antibody incubation: incubate the membrane in primary antibody solutions (rabbit anti-A3G, 1:2,500) overnight at 4°C (or in the cold room) on a rocking platform.

12. Washing: remove the primary antibody solution (which can be reused, see Note 11). Wash the membrane three to four times for 15–20 min each with 50 ml of washing buffer at room temperature on a rocking platform.

13. Secondary antibody incubation: incubate the membrane in freshly prepared secondary antibody solutions (1:5,000) for 1 h at room temperature on a rocking platform.

14. Washing: discard the secondary antibody solution. Wash the membrane three to four times for 15–20 min each with 50 ml of washing buffer at room temperature on a rocking platform.

15. During the final wash, aliquot 3 ml of each portion of the ECL reagent and warm them separately to room temperature. Once the final wash is done and the washing buffer is disregarded, mix the ECL reagents together and immediately add the mixed solution to the membrane and incubate for 1 min.

16. Remove the membrane from the ECL reagents, blot it with Kim-Wipes, and then place it between the leaves of an acetate sheet protector cut to the size of an X-ray film cassette and place it in an X-ray film cassette.

17. In a dark room under safe light conditions, place film in the X-ray film cassette and expose for a suitable time, typically 30 s to a few minutes. An example of the results produced is shown in Figs. 1 and 2.

18. Once a satisfactory exposure for the result of the A3G has been obtained, strip off the signal by incubating the membrane in warm stripping buffer (50 ml per blot) with agitation for 30 min at 70°C.

19. Wash the membrane extensively in washing buffer three to four times for 15–20 min each, using 100 ml for wash, on a rocking platform.

20. The membrane is now ready to be blocked again in blocking buffer and then reprobed with antibodies that recognize cofactors of the HMM A3G complexes. An example of the results is shown in Fig. 2b.

4. Notes

1. Invert the Ficoll-Paque Plus bottle several times to ensure thorough mixing before using. Do not use if a distinct yellow color or particulate material appears in the clear solution.

2. The erythrocytes are aggregated by the Ficoll and sediment through the Ficoll-Paque Plus to the bottom of the tubes. Because of their lower density, the lymphocytes are found at the interface between plasma and the Ficoll-Paque Plus with other slowly sedimenting particles, such as monocytes and platelets. The yield and degree of purity of the lymphocytes depend on the efficiency of red cell removal. When erythrocytes are aggregated, some lymphocytes are trapped in the clumps and sediment with the erythrocytes. This tendency can be reduced by diluting the buffy coat. Aggregation of erythrocytes is, however, enhanced at higher temperatures (37°C) which decreases yield, but at low temperatures (4°C) the rate of aggregation is decreased, increasing the time of separation. It is critical to perform the density-gradient centrifugation at 18–20°C, which gives optimum results.

3. Work fast, keep cells cold, and use pre-cooled solution. Higher temperature and longer incubation times may lead to nonspecific cell labeling.

4. Monocytes express low or medium level of CD4 and can be labeled with CD4 microbeads and copurified with CD4 T cells. It is critical to remove the residual monocytes by plastic adherence.

5. Do not exceed the indicated amount of A3G plasmid because too much overexpression of A3G in 293T cells will lead to the appearance of LMM form, suggesting there are limited cofactors required for the HMM assembly.

6. Use EDTA-free protease inhibitor because a high concentration of EDTA could disrupt the HMM complexes.

7. Perform all lysate preparation and protein fractionation steps at 4°C to prevent the HMM complexes from disassembling.

8. Perform all protein purification steps at 4°C to prevent the interacting proteins from dissociating.

9. Transfer a small aliquot of the supernatant to a Eppendorf tube. Store the aliquot at −30°C for later Western blot analysis (5 μl per lane is enough). It is really important, especially when performing the TAP expression/purification for the first time, to take aliquots for analysis of the purification procedures at all the steps indicated. Steps that do not yield efficient recovery should be troubleshot individually.

10. Keep a small aliquot of the beads before and after elution. Store the aliquots at –30°C for later Western blot analysis. It is really important to determine whether the binding and elution steps have been performed efficiently. Steps that do not yield efficient recovery should be troubleshot individually.

11. The primary antibody solution containing 0.02% sodium azide can be saved for subsequent experiments by storage at 4°C. It can be reused for up to 40 blots over 6 months. Often, the background gets cleaner as the hybridization mixture is reused. Filter through a 0.22-μm filter unit if particulate material appears in the clear solution.

References

1. Sheehy, A. M., Gaddis, N. C., Choi, J. D., and Malim, M. H. (2002) Isolation of a human gene that inhibits HIV-1 infection and is suppressed by the viral Vif protein. *Nature 418*, 646–650.

2. Chiu, Y. L. and Greene, W. C. (2008) The APOBEC3 cytidine deaminases: an innate defensive network opposing exogenous retroviruses and endogenous retroelements. *Annu Rev Immunol 26*, 317–353.

3. Soros, V. B., Yonemoto, W., and Greene, W. C. (2007) Newly synthesized APOBEC3G is incorporated into HIV virions, inhibited by HIV RNA, and subsequently activated by RNase H. *PLoS Pathog 3*, e15.

4. Bishop, K. N., Verma, M., Kim, E. Y., Wolinsky, S. M., and Malim, M. H. (2008) APOBEC3G inhibits elongation of HIV-1 reverse transcripts. *PLoS Pathog 4*, e1000231.

5. Guo, F., Cen, S., Niu, M., Saadatmand, J., and Kleiman, L. (2006) Inhibition of formula-primed reverse transcription by human APOBEC3G during human immunodeficiency virus type 1 replication. *J Virol 80*, 11710–11722.

6. Mangeat, B., Turelli, P., Caron, G., Friedli, M., Perrin, L., and Trono, D. (2003) Broad antiretroviral defence by human APOBEC3G through lethal editing of nascent reverse transcripts. *Nature 424*, 99–103.

7. Lecossier, D., Bouchonnet, F., Clavel, F., and Hance, A. J. (2003) Hypermutation of HIV-1 DNA in the absence of the Vif protein. *Science 300*, 1112.

8. Harris, R. S., Bishop, K. N., Sheehy, A. M., Craig, H. M., Petersen-Mahrt, S. K., Watt, I. N., Neuberger, M. S., and Malim, M. H. (2003) DNA deamination mediates innate immunity to retroviral infection. *Cell 113*, 803–809.

9. Zhang, H., Yang, B., Pomerantz, R. J., Zhang, C., Arunachalam, S. C., and Gao, L. (2003) The cytidine deaminase CEM15 induces hypermutation in newly synthesized HIV-1 DNA. *Nature 424*, 94–98.

10. Mbisa, J. L., Barr, R., Thomas, J. A., Vandegraaff, N., Dorweiler, I. J., Svarovskaia, E. S., Brown, W. L., Mansky, L. M., Gorelick, R. J., Harris, R. S., Engelman, A., and Pathak, V. K. (2007) HIV-1 cDNAs produced in the presence of APOBEC3G exhibit defects in plus-strand DNA transfer and integration. *J Virol 81*, 7099–7110.

11. Luo, K., Wang, T., Liu, B., Tian, C., Xiao, Z., Kappes, J., and Yu, X. F. (2007) Cytidine deaminases APOBEC3G and APOBEC3F interact with HIV-1 integrase and inhibit proviral DNA formation. *J Virol 81*, 7238–7248.

12. Stopak, K., de Noronha, C., Yonemoto, W., and Greene, W. C. (2003) HIV-1 Vif blocks the antiviral activity of APOBEC3G by impairing both its translation and intracellular stability. *Mol Cell 12*, 591–601.

13. Conticello, S. G., Harris, R. S., and Neuberger, M. S. (2003) The Vif protein of HIV triggers degradation of the human antiretroviral DNA deaminase APOBEC3G. *Curr Biol 13*, 2009–2013.

14. Marin, M., Rose, K. M., Kozak, S. L., and Kabat, D. (2003) HIV-1 Vif protein binds the editing enzyme APOBEC3G and induces its degradation. *Nat Med 9*, 1398–1403.

15. Sheehy, A. M., Gaddis, N. C., and Malim, M. H. (2003) The antiretroviral enzyme APOBEC3G is degraded by the proteasome in response to HIV-1 Vif. *Nat Med 9*, 1404–1407.

16. Yu, X., Yu, Y., Liu, B., Luo, K., Kong, W., Mao, P., and Yu, X. F. (2003) Induction of

APOBEC3G ubiquitination and degradation by an HIV-1 Vif-Cul5-SCF complex. *Science* **302**, 1056–1060.

17. Mariani, R., Chen, D., Schrofelbauer, B., Navarro, F., Konig, R., Bollman, B., Munk, C., Nymark-McMahon, H., and Landau, N. R. (2003) Species-specific exclusion of APOBEC3G from HIV-1 virions by Vif. *Cell* **114**, 21–31.

18. Chiu, Y. L., Soros, V. B., Kreisberg, J. F., Stopak, K., Yonemoto, W., and Greene, W. C. (2005) Cellular APOBEC3G restricts HIV-1 infection in resting CD4+ T cells. *Nature* **435**, 108–114.

19. Kreisberg, J. F., Yonemoto, W., and Greene, W. C. (2006) Endogenous factors enhance HIV infection of tissue naive CD4 T cells by stimulating high molecular mass APOBEC3G complex formation. *J Exp Med* **203**, 865–870.

20. Stopak, K. S., Chiu, Y. L., Kropp, J., Grant, R. M., and Greene, W. C. (2007) Distinct patterns of cytokine regulation of APOBEC3G expression and activity in primary lymphocytes, macrophages, and dendritic cells. *J Biol Chem* **282**, 3539–3546.

21. Ellery, P. J., Tippett, E., Chiu, Y. L., Paukovics, G., Cameron, P. U., Solomon, A., Lewin, S. R., Gorry, P. R., Jaworowski, A., Greene, W. C., Sonza, S., and Crowe, S. M. (2007) The CD16+ monocyte subset is more permissive to infection and preferentially harbors HIV-1 in vivo. *J Immunol* **178**, 6581–6589.

22. Pion, M., Granelli-Piperno, A., Mangeat, B., Stalder, R., Correa, R., Steinman, R. M., and Piguet, V. (2006) APOBEC3G/3F mediates intrinsic resistance of monocyte-derived dendritic cells to HIV-1 infection. *J Exp Med* **203**, 2887–2893.

23. Wang, F. X., Huang, J., Zhang, H., and Ma, X. (2008) APOBEC3G upregulation by alpha interferon restricts human immunodeficiency virus type 1 infection in human peripheral plasmacytoid dendritic cells. *J Gen Virol* **89**, 722–730.

24. Kamata, M., Nagaoka, Y., and Chen, I. S. (2009) Reassessing the role of APOBEC3G in human immunodeficiency virus type 1 infection of quiescent CD4+ T-cells. *PLoS Pathog* **5**, e1000342.

25. Santoni de Sio, F. R. and Trono, D. (2009) APOBEC3G-depleted resting CD4+ T cells remain refractory to HIV1 infection. *PLoS One* **4**, e6571.

26. Chen, K., Huang, J., Zhang, C., Huang, S., Nunnari, G., Wang, F. X., Tong, X., Gao, L., Nikisher, K., and Zhang, H. (2006) Alpha interferon potently enhances the anti-human immunodeficiency virus type 1 activity of APOBEC3G in resting primary CD4 T cells. *J Virol* **80**, 7645–7657.

27. Vetter, M. L. and D'Aquila, R. T. (2009) Cytoplasmic APOBEC3G restricts incoming Vif-positive HIV-1 and increases 2-LTR circle formation in activated T helper subtype cells. *J Virol* **83**, 8646–8654.

28. Lafferty, M. K., Sun, L., Demasi, L., Lu, W., and Garzino-Demo, A. CCR6 ligands inhibit HIV by inducing APOBEC3G. *Blood* **115**, 1564–1571.

29. Chiu, Y. L., Witkowska, H. E., Hall, S. C., Santiago, M., Soros, V. B., Esnault, C., Heidmann, T., and Greene, W. C. (2006) High-molecular-mass APOBEC3G complexes restrict Alu retrotransposition. *Proc Natl Acad Sci USA* **103**, 15588–15593.

30. Gallois-Montbrun, S., Kramer, B., Swanson, C. M., Byers, H., Lynham, S., Ward, M., and Malim, M. H. (2007) Antiviral protein APOBEC3G localizes to ribonucleoprotein complexes found in P bodies and stress granules. *J Virol* **81**, 2165–2178.

31. Kozak, S. L., Marin, M., Rose, K. M., Bystrom, C., and Kabat, D. (2006) The anti-HIV-1 editing enzyme APOBEC3G binds HIV-1 RNA and messenger RNAs that shuttle between polysomes and stress granules. *J Biol Chem* **281**, 29105–29119.

32. Wichroski, M. J., Robb, G. B., and Rana, T. M. (2006) Human retroviral host restriction factors APOBEC3G and APOBEC3F localize to mRNA processing bodies. *PLoS Pathog* **2**, e41.

33. Huang, J., Liang, Z., Yang, B., Tian, H., Ma, J., and Zhang, H. (2007) Derepression of microRNA-mediated protein translation inhibition by apolipoprotein B mRNA-editing enzyme catalytic polypeptide-like 3G (APOBEC3G) and its family members. *J Biol Chem* **282**, 33632–33640.

34. Dewannieux, M., Esnault, C., and Heidmann, T. (2003) LINE-mediated retrotransposition of marked Alu sequences. *Nat Genet* **35**, 41–48.

35. Kazazian, H. H. J. (2004) Mobile elements: drivers of genome evolution. *Science* **303**, 1626–1632.

36. Hulme, A. E., Bogerd, H. P., Cullen, B. R., and Moran, J. V. (2007) Selective inhibition of Alu retrotransposition by APOBEC3G. *Gene* **390**, 199–205.

37. Boyum, A. (1968) Isolation of mononuclear cells and granulocytes from human blood. Isolation of mononuclear cells by one centrifugation, and of granulocytes by combining centrifugation and sedimentation at 1 g. *Scand J Clin Lab Invest Suppl* **97**, 77–89.

38. Bain, B. and Pshyk, K. (1972) Enhanced reactivity in mixed leukocyte cultures after separation of mononuclear cells on Ficoll-Hypaque. *Transplant Proc* **4**, 163–164.

39. Rigaut, G., Shevchenko, A., Rutz, B., Wilm, M., Mann, M., and Seraphin, B. (1999) A generic protein purification method for protein complex characterization and proteome exploration. *Nat Biotechnol* **17**, 1030–1032.

40. Puig, O., Caspary, F., Rigaut, G., Rutz, B., Bouveret, E., Bragado-Nilsson, E., Wilm, M., and Seraphin, B. (2001) The tandem affinity purification (TAP) method: a general procedure of protein complex purification. *Methods* **24**, 218–229.

41. Wilson, D. S., Keefe, A. D., and Szostak, J. W. (2001) The use of mRNA display to select high-affinity protein-binding peptides. *Proc Natl Acad Sci USA* **98**, 3750–3755.

42. Keefe, A. D., Wilson, D. S., Seelig, B., and Szostak, J. W. (2001) One-step purification of recombinant proteins using a nanomolar-affinity streptavidin-binding peptide, the SBP-Tag. *Protein Expr Purif* **23**, 440–446.

43. Stofko-Hahn, R. E., Carr, D. W., and Scott, J. D. (1992) A single step purification for recombinant proteins. Characterization of a microtubule associated protein (MAP 2) fragment which associates with the type II cAMP-dependent protein kinase. *FEBS Lett* **302**, 274–278.

Part IV

tRNA Editing and RNA Modifications

Chapter 13

Analysis of tRNA Editing in Native and Synthetic Substrates

Jessica L. Spears, Kirk W. Gaston, and Juan D. Alfonzo

Abstract

The primary sequence of all nucleic acids in a cell contain 4 canonical nucleotides (G, A, T, and C for DNA and G, A, U, and C for RNA). However, post-transcriptionally, nucleic acids can undergo a number of chemical modifications, which may change their structure and function. tRNAs contain the most diverse array of post-transcriptionally added chemical groups that involve both editing and modification. Because editing and modification events can serve vital roles in cell function, it is important to develop techniques that allow for fast and accurate analysis of these events. This chapter describes the methods used to purify tRNAs from total native RNA pools and for subsequent analysis of their edited and modified states using reverse transcriptase-based approaches. These techniques, in combination with 2D-TLC, allow for the routine analysis and quantitation of edited and modified nucleotides in a fast, cost effective manner and without the need for special equipment such as HPLC or a mass spectrometer. Admittedly, the techniques described here are only applicable to a subset of post-transcriptional changes occurring in a tRNA such as C to U and A to I editing as well as modifications that prevent reverse transcriptase elongation; these have been highlighted throughout the chapter.

Key Words: tRNA editing, modification, native tRNA purification, poisoned primer extension

1. Introduction

Edited and modified nucleotides are common in nucleic acids belonging to all three domains of life. While the enzymes responsible for these reactions can recognize DNA and many types of RNAs (i.e. mRNA, rRNA, small non-coding RNAs, etc.), tRNAs are by far the most diversely and extensively modified substrates with over 100 modifications known to date.

Powerful methods such as mass spectrometry are gaining popularity and can be used to precisely detect, locate, and identify editing and modification events; however, these techniques require significant amounts of material for analysis, utilize expensive equipment and are often inaccessible to many researchers.

Ruslan Aphasizhev (ed.), *RNA and DNA Editing: Methods and Protocols*, Methods in Molecular Biology, vol. 718, DOI 10.1007/978-1-61779-018-8_13, © Springer Science+Business Media, LLC 2011

Therefore, methods more applicable to general laboratories are not only useful but also necessary for the routine quantitation and analysis of editing and modification. Grosjean et al. recently published an invaluable review of one such method involving the detection and quantification of modifications in tRNAs via two-dimensional thin layer chromatography (2D-TLC) (1). While 2D-TLC can be used to detect total modifications present in RNA, its use in the actual mapping and assignment of single modifications to specific locations in a nucleic acid is technically challenging. The present report focuses on techniques used for the analysis of editing and modification at a specific position within native tRNAs.

Routinely, we begin our analysis by performing RT–PCR using tRNA-specific oligonucleotides and total RNA from our organism(s) of choice (trypanosomes). By cloning the PCR product and subsequently sequencing the clones, we are able to establish the presence of editing events. If editing is detected, the levels of editing are further quantitated using "poisoned" primer extension (PPE) (Fig. 1). The PPE protocol uses total native RNA from trypanosomatids and tRNA-specific ^{32}P-labeled oligonucleotides to locate editing events via reverse transcription. In this modified

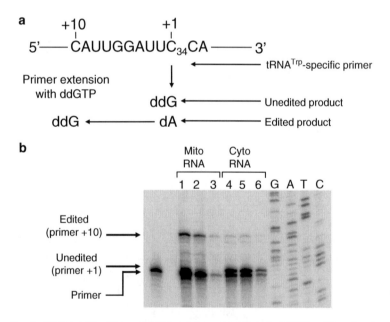

Fig. 1. **"Poisoned" primer extension assay to determine tRNA editing levels.** **(a)** Schematic representation of the assay, ddGTP refers to the guanosine analog dideoxyGTP used to stop ("poison") the primer extension reaction. C_{34} indicates the edited position. In the case of the unedited tRNA (C_{34}) ddGTP is incorporated and the extension reaction terminates. In the case of an edited tRNA (U_{34}) dATP is incorporated and thus the primer extension continues until 9 nucleotides later when ddGTP is incorporated. **(b)** Assay quantitation using cytosolic (cyto) and mitochondrial (mito) tRNA^{Trp}. The primer and the edited and unedited products are indicated.

version of a primer extension, the reaction includes any three deoxynucleotide triphosphates but the fourth nucleotide is replaced for its dideoxynucleotide analog specific for the edited position (2). For example, to detect a C to U change the reaction is performed in the presence of dA, dT, dC, and ddG. In this reaction reverse transcriptase incorporates the chain terminator ddG across from the edited position yielding two possible products: a primer extended up to the unedited based and a second product where editing prevents ddG incorporation and the reaction then stops at the next available C in the RNA sequence. The amount of both products (edited and non-edited) can be quantitated and compared. Similar reactions can be performed for the detection of A to I editing where ddG is then replaced by ddT. Since inosine is a guanosine analog, in this case primer extension will stop at the unedited position following ddT incorporation.

In the presence of a modification that blocks reverse transcriptase (i.e. a modification on the Watson–Crick face of the nucleotide, m^3C or m^1A for example) even if all four deoxynucleotides are present, the primer extension reaction will end one nucleotide short of the modified position (Fig. 2). While primer extension is not sufficient to assess the chemical nature of the modification(s), it does confirm the presence of a modification that inhibits base pairing. In these cases, we enrich for tRNAs

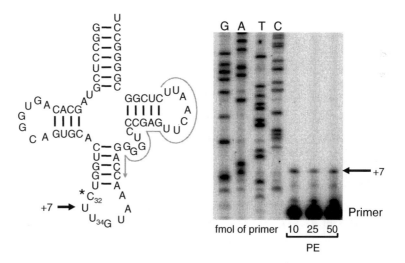

Fig. 2. **Primer extension of tRNA$^{Thr}_{UGU}$ from Trypanosoma brucei.** Primer extensions (PE) were performed using total RNA from *Trypanosoma brucei* and a 5′ labeled oligonucleotide specific for tRNA$^{Thr}_{UGU}$. The oligonucleotide primer is as indicated by the gray arrow and the position of the product is indicated by a black arrow with the number of nucleotides extended noted. A sequencing ladder (seen in the first 4 lanes) aided in determining the length of the extended product. The asterisk denotes the presence of a modification that prevents reverse transcriptase elongation ("hard stop"). In this particular case the modification has now been confirmed by mass spectrometry to be m^3C/m^3U (Gaston and Alfonzo, unpublished results).

(from total RNA) and further affinity purify a single tRNA species. The tRNA is post-labeled and the total modifications are analyzed by 2D-TLC followed by liquid chromatography and tandem mass spectrometry (LC/MS/MS). We are also able to explore more mechanistic aspects of a given reaction using tRNA transcripts and one-dimensional thin layer chromatography (1D-TLC). Here cell extracts or purified enzymes are incubated with a radio-labeled tRNA transcript and the editing or modification activity of the extract/purified enzyme can be followed by 1D-TLC.

In the present chapter we focus on techniques routinely used in our laboratory to detect and quantitate editing and modifications. While these methods are not applicable to every modification, they can be used for the study of two types of reactions: (1) those that exchange one canonical nucleotide for another (editing) and (2) those that occur on the Watson and Crick edge of the base, posing a roadblock for reverse transcriptase and thus producing a "hard" stop in a primer extension assay (Fig. 2). We will also describe protocols for the enrichment of small RNAs from total RNA including the use of biotinylated oligonucleotides for the affinity purification of single tRNA species which can then be analyzed for modification content by 2D-TLC. More elaborate techniques such as mass spectrometry are beyond our expertise and will only be mentioned briefly.

2. Materials

All chemicals used are of reagent grade or higher. Double-distilled water (ddH$_2$O) is made RNase-free by autoclaving. Gloves are worn throughout all procedures to minimize possible RNase contamination. Materials and chemicals needed are listed under subheadings which match those subheadings seen in the methods section. Buffer components are listed at final concentration. The manufacturers of critical enzymes and reagents are noted.

2.1. tRNA Editing Analysis by RT–PCR

1. 5× 1st strand buffer as supplied with Superscript III reverse transcriptase (Invitrogen).
2. tRNA-specific oligonucleotide (40 µM stock).
3. 10 mM dNTPs (2.5 mM of each dNTP).
4. Superscript III reverse transcriptase (Invitrogen).
5. 0.1 M dithiothreitol (DTT).
6. PCR reagents. We typically use 1% NP40, 50% acetamide, 10× PCR buffer, 12.5 mM dNTPs and Taq polymerase in our PCR reactions.

2.2. tRNA Editing Analysis by 3' Poly(A)-tailing and RT–PCR

1. 5× Yeast poly(A) polymerase (PAP) reaction buffer as supplied with the polymerase enzyme (USB/Affymetrix).
2. 0.1 M dithiothreitol (DTT).
3. 0.1 M adenosine triphosphate (ATP).
4. Yeast PAP (USB/Affymetrix).
5. 5× *E. coli* PAP supplement buffer as supplied with the polymerase enzyme (Invitrogen).
6. 0.1 M MnCl$_2$.
7. *E. coli* Poly A Polymerase (Invitrogen).
8. Phenol saturated with Tris–HCl, pH 8.0.
9. Ethanol.
10. 3M sodium acetate, pH 5.2 (NaOAc).
11. Glycogen (10 mg/ml) (used as a carrier during ethanol precipitation steps).

2.3. tRNA Editing Analysis by End-Tagging and RT–PCR

1. Agarose.
2. Ethidium bromide (EtBr) (5 mg/ml).
3. PAGE buffer (10× TBE): 0.9M Tris, 0.9 M Boric acid, 20 mM EDTA, pH 8.0.
4. Phenol saturated with Tris–HCl, pH 8.0.
5. T4 RNA ligase 1 (New England Biolabs).
6. T4 RNA ligase 1 buffer as supplied with the enzyme (New England Biolabs).
7. Anchor ligation oligonucleotide (40 µM stock): 5'-pTCTT-TAGTGAGGGTTAATTGCCddC-3' (see Note 1).
8. Ethanol.
9. 3 M sodium acetate (NaOAc pH 5.2).
10. Glycogen (10 mg/ml) (used as a carrier during ethanol precipitation steps).

2.4. Analysis of tRNA Modifications by Primer Extension

1. Hybridization buffer: 30 mM Tris–HCl, pH 8.3, 0.45 M KCl.
2. [γ-^{32}P] ATP (6000 Ci/mmol)(MP Biomedicals).
3. T4 polynucleotide kinase (PNK) (Invitrogen).
4. 0.5 M ethylenediaminetetraacetic acid (EDTA).
5. 2 M NaOH.
6. 25 mM MgCl$_2$.
7. 2.5 mM dNTPs and 1.25 mM ddNTPs (see Note 2).
8. 1 M Tris–HCl, pH 8.3.
9. 0.1 M dithiothreitol (DTT).

10. Ribonuclease inhibitor (RNasin) (Promega, Madison, WI).

11. Superscript III reverse transcriptase (Invitrogen).

12. tRNA-specific oligonucleotides (40 μM.stock).

13. 15% acrylamide/bis solution, 8 M urea, N,N,N,N'-tetramethyl-ethylenediamine (TEMED), and 10% ammonium persulfate (APS).

14. PAGE buffer (10× TBE): 0.9 M Tris, 0.9 M boric acid, 20 mM EDTA, pH 8.0.

15. Urea loading buffer: 7 M urea in TE buffer (10 mM Tris–HCl, pH 7.5, 1 mM EDTA).

2.5. Enrichment and Purification of tRNA

1. Qiagen-tip 2500 column (Qiagen) (see Note 3).

2. Equilibration buffer: 50 mM 3-(N-Morpholino) propanesulfonic acid (MOPS), pH 7.0, 15% Isopropanol, 1% Triton X-100.

3. Wash buffer: 50 mM MOPS, pH 7.0, 0.2 M NaCl.

4. Elution buffer: 50 mM MOPS, pH 7.0, 0.75 M NaCl, 15% ethanol.

5. 1 M MOPS.

6. Isopropanol.

7. Ethanol.

8. TE buffer: 10 mM Tris–HCl, pH 7.5, 1 mM ethylenediaminetetraacetic acid (EDTA).

9. Streptavidin agarose beads (Pierce, Rockford, IL).

10. 20× SSC, pH 7.0, (3 M NaCl, 0.3 M Sodium citrate).

11. 40% polyacrylamide/bis solution (39:1), 8 M urea, N,N,N,N'-tetramethyl-ethylenediamine (TEMED), and 10% ammonium persulfate (APS).

12. PAGE buffer (10× TBE): 0.9 M Tris, 0.9 M boric acid, 20 mM EDTA, pH 8.0.

13. Acrylamide gel elution buffer (0.3 M sodium acetate, pH 5.2).

14. 1% sodium dodecyl sulfate (SDS).

2.6. Analysis of the Modification Content of Native tRNAs

1. RNaseT2 (Sigma).

2. RNaseT2 buffer as supplied with RNase T2 (Sigma).

3. 10× kinase buffer as supplied with T4 polynucleotide kinase (Invitrogen).

4. 10× BSA (10 mg/ml).

5. T4 polynucleotide kinase (PNK) (Invitrogen).

6. [γ-^{32}P] ATP (6000 Ci/mmol) (MP Biomedicals).

7. 0.2 M glycerol.

8. Glycerol kinase ("Glycerokinase" by Roche).

9. 30 mM ATP.

10. Chloroform.

11. Water-saturated ether.

2.7. In vitro *Analysis of tRNA Modification Enzyme Activity*

1. [α-^{32}P]-NTP (MP Biomedicals).

2. 5× Reaction buffer optimized for the enzyme being assayed. For editing deaminases, this buffer contains 0.25 M Tris–HCl, pH 8.0, a source of Mg^{2+} (i.e. $MgCl_2$ or $MgSO_4$), and 0.5 mM EDTA. In the case of the *T. brucei* ADAT2/3 deaminase, 10 mM DTT must be added for enzyme activity.

3. 10× nuclease P1 buffer as supplied with nuclease P1 (MP Biomedicals).

4. Nuclease P1 (MP Biomedicals).

3. Methods

3.1. tRNA Editing Analysis by RT–PCR

To analyze the editing levels of tRNAs in a total native RNA population, a cDNA copy of the tRNA is made using reverse transcriptase (3, 4) and a tRNA-specific primer that anneals to the 3′ end of the tRNA. A subsequent PCR is performed using the cDNA as template and two tRNA-specific oligonucleotides (that anneal to the 5′ and 3′ end, respectively). The resulting PCR product is cloned into a pCR2.1TOPO-TA plasmid vector and ~30 clones are sequenced.

If editing events of interest are located near the 5′ or 3′ end of the tRNA, i.e. covered by the oligonucleotides, it is possible to add a 3′ tail (see 3.2) and/or 5′ tag (see 3.3) to the tRNA so those editing events can be analyzed.

1. Assemble two reactions for each tRNA sample. One reaction will serve as the RT– control (label tube as such). The other reaction will be the RT+ test. Add 2 pmol of the 3′ tRNA-specific primer to 5 µg of total RNA (see Note 4). Use water to increase the volume of this reaction to 12 µL.

2. Incubate the reaction at 70°C for 10 minutes and then chill on ice for 1 minute.

3. To each reaction add 4 µL of 5× 1st strand buffer, 2 µL 0.1 M DTT, and 1 µL dNTPs (10 mM each). Mix and incubate at 42°C for 2 minutes.

4. Add 1 µL of Superscript III reverse transcriptase to each RT+ reaction tube. Remember that 1 tube (RT–) in each set will not receive reverse transcriptase and will serve as a negative control.

5. Incubate all reactions (RT+ and RT–) at 42°C for 50 minutes.

6. Perform a standard PCR reaction with cycling conditions optimized for the oligonucleotides being used. Use a forward 5′ oligonucleotide specific for the tRNA of interest and the same 3′ reverse oligonucleotide used in the RT reaction.

7. The resulting RT–PCR product is further cloned into pCR2.1-TOPO plasmid vector per manufacturer's instructions, transformed into *E. coli* and independent clones are sequenced. The sequences obtained will be representative of the number of tRNAs that undergo adenosine to inosine or cytidine to uridine editing *in vivo*. We typically sequence at least 30 independent clones to ensure our edited to unedited tRNA ratios are statistically significant.

3.2. tRNA Editing Analysis by 3′ Poly(A)-tailing and RT–PCR

tRNAs can be tagged by 3′ poly(A) tailing. A poly(A) tail is added to the tRNA by two different poly A polymerase reactions. The first reaction utilizes yeast poly(A) polymerase while the second reaction utilizes *E. coli* poly(A) polymerase. The tailed tRNAs are reverse transcribed (3, 4) using an oligo-T oligonucleotide (20mer with 18 Ts and 2 Gs) and then amplified by PCR. The resulting PCR product (tRNA cDNA) is TOPO-cloned as above (5) and the clones are sequenced to determine the level of editing in a given tRNA.

1. Assemble the first poly(A) reaction by adding 20 μL of 5× yeast PAP buffer (USB), 1 μL of 0.1 M DTT, 0.5 μL of 0.1 M ATP, 2 μL of yeast PAP (USB) and tRNA (10–150 ng) to a 1.5 mL microcentrifuge tube. Use water to bring the total reaction volume up to 100 μL.

2. Incubate the reaction at 37°C for 45 minutes.

3. To each reaction, add 30 μL 5× *E. coli* PAP supplement buffer, 7.5 μL of 10 mM ATP, 3.5 μL of 0.1 M MnCl$_2$, and 2 μL of *E. coli* Poly A Polymerase.

4. Incubate this reaction at 37°C for 45 minutes.

5. Extract the tRNA with 100 μL of Tris-saturated phenol and spin the sample in a microcentrifuge at $16,000 \times g$ for 10 min at 4°C.

6. Carefully transfer the top aqueous phase to a clean microcentrifuge tube.

7. Precipitate the tRNA by adding 3 volumes of 100% ethanol, 1/10th volume of 3 M NaOAc, and 2 μL glycogen. Spin the sample in a microcentrifuge at $16,000 \times g$ for 30 minutes at 4°C.

8. Remove the supernatant by pipetting.

9. Resuspend the pellet in a minimal amount of ddH$_2$O. Store at –20°C.

10. The RT and PCR reactions are carried out as above (3.1) with the following changes: the 3′ reverse oligonucleotide used in the RT reaction as well as the PCR is an oligo-T oligonucleotide. The 5′ forward oligonucleotide is tRNA specific or specific for the 5′ tag (see 3.3).

3.3. tRNA Editing Analysis by End-Tagging and RT–PCR

In case of an editing event occurring near the 5′ end of a tRNA, it is necessary to tag the tRNA. After the reverse transcription reaction a tagging oligonucleotide can be ligated to the end of the cDNA; a PCR is then performed with oligonucleotides specific for the tag added to the end of the cDNA copy (6). The entire tRNA can be cloned and sequenced by this procedure, thus solving any issues of the oligonucleotides used to amplify the RNA occluding editing sites.

1. Purify the large cDNAs from the reverse transcription reaction described above (3.1 or 3.2) using a 3% agarose gel stained with ethidium bromide or other appropriate dye. The largest of the fragments (tRNA copy plus the two end tags, 70–150nt) will constitute the DNA fragment corresponding to the full-length tRNA.

2. Excise the full-length DNA fragment from the gel with a sterile razor blade. Place the gel slice in a fresh 2 mL microcentrifuge tube. To the gel slice, add an equal volume of Tris-saturated phenol, pH 8.0, and incubate it at –20°C for 1 hour to overnight to completely freeze the phenol/gel slice.

3. Thaw the phenol/gel slice mixture and spin in a microcentrifuge at 16,000 × *g* for 15 minutes. Transfer the aqueous phase to a new tube and precipitate the sample by adding 2.5 volumes of ethanol, 1/10th volume of 3 M NaOAc (pH 5.2), and 1 μL of glycogen. Spin the sample in a microcentrifuge at 16,000 × *g* for 30 minutes.

4. Remove the ethanol and resuspend the pellet in 43 μL of water. Add 1 μL of the anchor ligation oligonucleotide (see Note 1), 5 μL of 10× T4 RNA ligase buffer, and 1 μL of T4 RNA ligase (total volume 50 μL). Incubate this reaction at room temperature for 16 hours and recover the DNA by ethanol precipitation. Dissolve the resulting pellet in 10 μL of water before proceeding with the PCR amplification procedure.

5. Follow 3.1, steps 6 and 7, but instead of tRNA-specific oligonucleotide, use the tag-specific oligonucleotide as the forward primer.

3.4. Analysis of tRNA Modifications by Primer Extension

1. Label 5 μmol of specific oligonucleotide with γ^{32}P-ATP using 1 μl of T4-PNK (polynucleotide kinase) in a 10 μl reaction volume per manufacturer's instruction. Purify the labeled oligonucleotide by passing it through a G-25 column.

2. Add 0.5–2.5 μmol of labeled oligonucleotide to 5–25 μg of RNA. Adjust the final volume to 50 μL with water and precipitate with 3 volumes of ethanol, 1/10th volume (5 μL) of 3 M NaOAc, and 1 μL of glycogen. Store the precipitation reaction in the freezer for 30 minutes to overnight and spin at $16,000 \times g$ for 30 minutes in a microcentrifuge.

3. Carefully remove the supernatant (see Note 5) and resuspend the pellet in 10 μL of water.

4. Add 5 μL of PE hybridization buffer to the RNA and heat at 90°C for 5 minutes. Transfer the reaction to a 37°C water bath and incubate for 90 minutes.

5. During the 90 minute incubation, mix the following buffers and reagents per reaction: 6.15 μL of water, 3.6 μL of 25 mM $MgCl_2$, 3.0 μL of 10 mM dNTPs, 1.35 μL of 1 M Tris–HCl, pH 8.3, 0.3 μL of 100 mM DTT, 0.1 μL of RNase inhibitor, and 0.5 μL of Superscript III. To perform poisoned primer extension one or more dNTPs should be left out of the reaction and substituted with ddNTPs (see Note 6).

6. After the 90 minute incubation, place both the hybridization reaction (step 4) and the reagent mix (step 5) at 42°C for 5 minutes. Add the contents of the reagent mix to the hybridization reaction (15 μl total per reaction) and incubate at 42°C for 1 hour.

7. After incubation, add 1.5 μL of 0.5 M EDTA and 3.2 μL of 2 M NaOH and incubate at 65°C for 30 minutes.

8. After incubation, add 3.2 μL of 2 M HCl and precipitate the cDNA by adding 1 μL of glycogen, 10 μL of 7.5 M NH_4OAc, and 150 μL of ethanol and placing the reaction at –20°C for 30 minutes to overnight. Spin the sample at $16,000 \times g$ for 30 minutes, remove the supernatant, wash the pellet with 70% ethanol, and dry in a speed-vac with heat (80°C).

9. Resuspend the pellet in 10 μL of urea loading buffer with dye. Heat sample at 85°C for 5 minutes and load 2–3 μL of sample per lane on a 10% acrylamide 8M urea gel (see Note 7).

10. After gel electrophoresis, place the gel on filter paper, cover in plastic wrap and vacuum-dry the gel.

11. Expose the gel to a PhosporImager screen (Figs. 1 and 2).

3.5. Enrichment and Purification of Native tRNAs

This subheading describes the methods used to first enrich tRNAs (and other small RNAs) from a pool of nucleic acids and subsequently purify a specific tRNA from that enriched pool (7, 8).

The following protocol begins with nucleic acids dissolved in water. We use an acid phenol-guanidinium thiocyanate-chloroform extraction method (9), although other protocols may be more appropriate depending on the organism from which the isolation is occurring.

The following tRNA enrichment protocol has been optimized for use with the Qiagen Tip 2500 column (Dr. Michael Ibba, personal communication). The tRNAs purified using the following protocol can be used for liquid chromatography and mass spectrometry analysis. These tRNAs are purified under native conditions and can also be post-labeled to analyze the modified nucleotide composition by two-dimensional thin layer chromatography (2D-TLC) as described here and in great detail by H. Grosjean (1).

1. Equilibrate a Qiagen Q-2500 column with 2× 50 mL of fresh equilibration buffer (see Note 8).

2. To the nucleic acid pool, add 1 M MOPS, pH 7.0, to a final concentration of 0.1 M.

3. Load the nucleic acids in 0.1 M MOPS onto the equilibrated Q-2500 column and collect the flow-through (see Note 9).

4. Wash the column with 200 mL of wash buffer by gravity flow.

5. Further wash the column with 10 mL of elution buffer.

6. Elute the RNA from the column with 20 mL of elution buffer.

7. Precipitate the eluted RNA by adding equal volume of isopropanol. Store the RNA in isopropanol at −20°C for 30 minutes to overnight before centrifugation. The RNA can be stored at −20°C in isopropanol indefinitely.

8. Spin the sample in a microcentrifuge at $16,000 \times g$ for 30 minutes, decant the supernatant, and wash the pellet with 70% ethanol to remove residual salts.

9. Resuspend the tRNA enriched RNA in a minimal volume of TE (10 mM Tris–HCl, pH 8.0, 1 mM EDTA) or water.

10. Wash 250 μL of Streptavidin agarose beads in a 2 mL microcentrifuge tube by resuspending the beads in 1.75 mL 6× SSC. Spin the beads at $16,000 \times g$ in a microcentrifuge for 15 minutes and remove the supernatant by pipetting. Repeat washing step at least 3 times. Resuspend the beads in 0.75 mL of 6× SSC.

11. Heat 20 μg of a biotinylated oligonucleotide (specific for the desired tRNA (see Note 10)) at 70°C for 5 minutes.

12. Add the heated oligonucleotide to the washed beads in 6× SSC.

13. Incubate the beads and oligonucleotide at room temperature for 1 hour.

14. After incubation spin the beads at $16,000 \times g$ for 15 minutes and remove the supernatant. Wash the beads twice with 6× SSC to remove any unbound oligonucleotide. At this point the tRNA-specific oligonucleotide is now bound to the beads.

15. Prepare the enriched RNA (from 3.5 Step 9) for hybridization to the oligonucleotide-bound beads by heating the RNA at 85°C for 10 minutes.

16. Lower the temperature of the RNA to 70°C and then adjust the final concentration of the mixture to 6× SSC using a 20× SSC stock.

17. Add the oligonucleotide-bound beads (from Step 14) to the RNA and incubate at 70°C for 10 minutes. Slowly cool the mixture to 37°C in a water bath. Hybridize the RNA to the oligonucleotide-bound beads for 2 hours at 37°C.

18. Wash the beads as done previously this time with 3× SSC at room temperature. Repeat this wash step 3 times to remove unbound RNA (see Note 11).

19. Elute the purified tRNA with 500 μL of elution buffer (0.1× SSC, 0.1% SDS) by heating the beads at 80°C for 5 minutes, spinning the sample at $16,000 \times g$ and collecting the elution (supernatant). Repeat this elution process 10 times (or 10 × 500 μl aliquots).

20. Precipitate the eluted RNA by adding 3 volumes of ethanol, 1/10th volume of 3 M NaOAc, and 2 μL glycogen and spin the samples in a microcentrifuge at $16,000 \times g$ for 30 minutes. Resuspend the pellet in 10 μL of water and then add 10 μL of denaturing loading dye (containing 7M urea).

21. Heat the sample at 80°C for 5 minutes and load the RNA onto an 8 M urea 15% polyacrylamide gel (18 cm length). Run the gel at 600 volts until the faster migrating dye (bromophenol blue) is near the bottom of the gel.

22. Visualize the RNA by placing the gel on clear plastic wrap; place plastic wrap on a PEI-cellulose UV-shadow plate. To visualize the RNA, expose the gel to UV light (245 nm). Excise the corresponding tRNA band from the gel using a sterile razor blade and place it in a microcentrifuge tube. Elute the tRNA from the gel by adding enough 0.3 M NaOAc to cover the entire gel slice and allow it to incubate at room temperature or 4°C for at least 2 hours to overnight (for typical results see Fig. 3).

3.6. 2D-TLC Analysis of Natively Purified tRNAs

Natively purified tRNAs can be post-labeled to analyze the modified nucleotide composition by two-dimensional thin layer chromatography (2D-TLC) (1). Published 2D-TLC maps can be

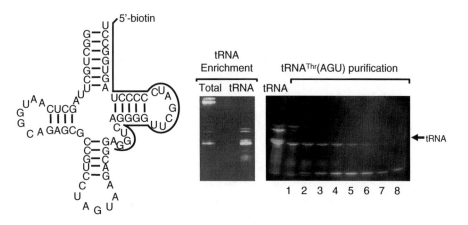

Fig. 3. **Purification of tRNA^{Thr}AGU.** Total native *Leishmania tarentolae* RNA was prepared and small RNAs were enriched from the total RNA. The enriched RNA was run on a 6% denaturing polyacrylamide gel and then stained with ethidium bromide. The enriched tRNA was then used to purify a single tRNA species, tRNA^{Thr}AGU, with a biotinylated oligonucleotide specific for this isoacceptor (the oligonucleotide is depicted on the tRNA 2D structure as a black line). To assess the purity of the RNA, the enriched total tRNA and the tRNA^{Thr}AGU) elutions (1–8) were run on a 10% denaturing polyacrylamide gel stained and stained ethidium bromide. tRNA^{Thr}AGU was then excised and eluted from the gel and further used in the modification analysis by methods described.

found in an earlier series chapter by H. Grosjean (1) or modified control RNA can be used to determine the migration of a new modification and complement the already published maps.

1. Digest tRNA with RNase T2 per manufacturer's protocol to yield 3′-monophosphates. Typically this digest is done in a 10 µL reaction with 100 ng to 1 µg of RNA.

2. Vacuum-dry the reaction on high heat.

3. To post-label every nucleotide, the digested tRNA is phosphorylated with γ^{32}P-ATP and polynucleotide kinase. Resuspend the dried pellet in the kinase reaction buffer: 7µL of ddH$_2$O, 1.5 µL of 10× kinase buffer, 4 µL of γ^{32}P-ATP, 1.5 µL of 10× BSA, and 1 µL of kinase. Incubate the reaction at 37°C for 30–180 minutes.

4. After the kinase reaction, inactivate the residual γ^{32}P-ATP by glycerol kinase: treat the 5′-labeled nucleotides by adding 3 µL of 0.2 M glycerol, 1 µL of glycerol kinase, and 23 µL of ddH$_2$O. Incubate the reaction at 37°C for 30 minutes. Add 1 µL of 30 mM ATP and incubate for an additional 30 minutes. Add an additional 1 µL of 30 mM ATP and incubate at 37°C again for 30 minutes. Finally the reaction is extracted with chloroform and re-extracted with water-saturated ether (the ether phase is the top phase). Vacuum-dry the sample to remove residual ether.

5. To cleave the 3′ phosphate, digest the labeled nucleotides with nuclease P1. Resuspend the dried pellet in 9 µL of

1× nuclease P1 buffer and add 1 μL of nuclease P1. Incubate the reaction at 37°C for at least 6 hours.

6. Vacuum-dry the reaction and resuspend the pellet in 5–10 μL of ddH$_2$O and spot 1 μL (100,000 cpm) on the bottom left corner of the TLC plate (the origin).

7. Place the TLC plate in a tank filled 5 mm deep with an appropriate solvent (see Note 12).

8. Allow the solvent front to migrate to the top of the plate (within 5 mm of the top edge of the plate). Remove the plate from the tank and allow it to completely air dry.

9. Rotate the plate 90° counterclockwise with respect to the origin and place the TLC in a tank filled 5 mm deep with the second solvent.

10. Allow the TLC to air dry and expose it to film or a PhosphorImager screen for visualization.

3.7. In vitro *Analysis of tRNA Modification Activity*

1. Radiolabel a tRNA substrate with α^{32}P-NTP by performing the T7 polymerase transcription reaction in the presence of a labeled NTP (see Note 13).

2. Denature 1 μL of the labeled tRNA per reaction by heating at 70°C for 3 minutes. Allow the tRNA to refold by slow cooling the tRNA at room temperature for 2 minutes. Add 1/10th the volume of 20 mM MgCl$_2$ (Mg^{2+} will help the cooled tRNA fold properly (see Note 14)).

3. Assemble a 50 μL reaction by adding 10 μL of 5× buffer and an appropriate amount of protein (or extract) of choice. Use water to bring the volume of the reaction up to 50 μL. Place the reaction at the optimal temperature for the organism from which the protein or extract is derived (see Note 15). Incubate the reaction at the appropriate temperature for 1 hour and then place on ice.

4. Phenol extract the RNA by adding 50 μL of Tris-phenol, pH 8.0, and spin the sample at 16,000 × g for 10 minutes in a microcentrifuge to separate the phenol and aqueous phase. Pipette out the aqueous phase (making sure not to disturb the phenol phase) to a clean microcentrifuge tube.

5. Precipitate the extracted RNA by adding 3 volumes of 100% ethanol, 1/10th volume 3 M NaOAc, and 2 μL of glycogen and spinning the sample at 16,000 × g for 30 minutes. Remove the ethanol and allow pellet to air dry.

6. To digest the tRNA into 5′-monophosphates, resuspend the dried pellet in 9 μL of 1× nuclease P1 buffer and 1 μL nuclease P1 and incubate the reaction at 37°C for at least 5 hours (see Note 16).

7. Dry the 10 μL digest reaction in a vacuum centrifuge. Resuspend the pellet in 2 μL of water. Spot 1 μL of this sample at the bottom of a silica thin layer chromatography plate.

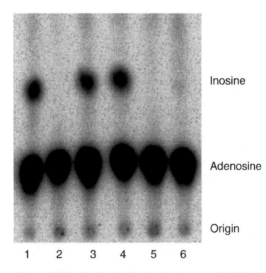

Fig. 4. **Adenosine deamination by Trypansoma brucei ADAT2/3 (Sample TLC).** α^{32}P-ATP labeled tRNA$^{Val}_{AAC}$ was incubated at 28°C in the presence of wild type or mutant TbADAT2/3 proteins for 1 hour. The reactions were then phenol extracted, ethanol precipitated, and digested overnight with nuclease P1. The resulting 5′ monophosphates were spotted and resolved on a TLC plate. Cold markers confirmed the adenosine and inosine spots as labeled. Lane 1, Wild type positive control; Lane 2, Negative control in which no protein was added to the reaction; Lanes 3 and 4 show two mutant proteins that are still able to deaminate the tRNA substrate; Lane 5 and 6 show two mutant proteins that are no longer able to deaminate the tRNA substrate.

This spot is the origin. Also spot an appropriate cold nucleotide marker (G, A, T, and/or C). Be sure the sample is at least 7 mm from the bottom edge of the plate. Allow sample to dry and place the plate in a tank filled 5 mm deep with an appropriate solvent (see Note 17).

8. Allow the solvent front to migrate to the top of the plate (within 5 mm of the top edge of the plate). Remove the plate from the tank and allow it to air dry. Wrap the TLC plate in plastic wrap and expose to a PhosphorImager screen. A sample with high specific activity can be visualized after 15–30 minutes but for best results, expose the TLC to the screen overnight. For sample results see Fig. 4.

9. Compare the results to cold nucleotide markers or published TLC maps (1).

4. Notes

1. Oligonucleotide sequence: 5′-pTCTTTAGTGAGGGTTAATT GCCddC-3′
   The 5′-p denotes a phosphorylated oligonucleotide so that the ligation can proceed. The ddC-3′ is a dideoxycytidine

added to prevent the 3′ end of the oligonucleotide from ligating to the 5′ end of the RNA.

2. In the case of detecting C to U editing use 2.5 mM dATP, dCTP, dTTP and 1.25 mM ddGTP. When detecting A to I editing, use 2.5 mM dATP, dCTP, dGTP, and 1.25 mM ddTTP.

3. The protocol has been optimized for this column. Columns of other sizes may be used although volumes will need to be adjusted. Brands other than Qiagen may work if the column resin is a similar anion exchange matrix but other brands have not been tested.

4. Typically the total RNA is DNase-treated before being used in this protocol.

5. A Geiger counter should be used to ensure that the pellet (labeled oligonucleotide) is not accidentally pipetted and removed.

6. To maximize the quality of the data, it is necessary to titrate the amount of the RNA in the reaction to find the optimal RNA to primer ratio. Use the ratio that yields at least half of the signal in the extension product.

7. A higher percentage acrylamide gel (i.e. 10–15%) will allow for better separation of those nucleotides near the primer.

8. The Qiagen-tip 2500 can be reused for subsequent tRNA purifications. After final tRNA elution, wash the column via gravity flow with 1× 50 mL of elution buffer followed by 1× 50 mL of ddH$_2$O. Cover the empty column with parafilm and store at 4°C. Before the next use, follow the equilibration step as written in 3.5 Step 1.

9. This step is completed on the bench top at room temperature; however, the flow through, washes, and elutions are collected on ice. Loading the flow through back onto the column at least 2 more times will maximize binding and increase the final yield of enriched tRNA. The flow through may be saved by freezing it at –20°C.

10. The oligonucleotide is typically 30 nucleotides long and should anneal to enough of the tRNA to make it specific for a single isoacceptor.

11. A 1× SSC wash (repeated twice) maybe added for increased stringency.

12. Usually we develop the first dimension of the TLC in a solvent containing isobutyric acid (66%), concentrated ammonia (1%), and water (33%) (by volume). There are two different solvents routinely used for developing the second dimension; one containing phosphate buffer, ammonium sulfate, and n-propanol in a 100:60:2 (v:w:v) ratio and the other containing

68% isopropanol (68%), concentrated HCl (18%) and water (14%) (by volume).

13. The radiolabeled nucleotide used will depend on the nucleotide that is potentially modified or edited. In the case of looking for adenosine to inosine deamination α^{32}P-ATP is used to uniformly label all the adenosines in the tRNA.

14. Heating RNA in the presence of Mg^{2+} can cleave the RNA. It is important that the tRNA be heated to 70°C in the absence of $MgCl_2$. The $MgCl_2$ can be added to the tRNA to assist in proper folding after the tRNA has been allowed to cool.

15. For example, assays containing *Trypanosoma brucei* ADAT2/3 protein over-expressed in and purified from *E. coli* are incubated at 28°C because this is optimal growth temperature for *T. brucei*. Assays performed with the Human ADAT2/3 protein over-expressed in and purified from *E. coli* are incubated at 37°C.

16. Digest the tRNA with nuclease P1 for at least 5 hours to ensure complete digestion. This digest reaction can be incubated overnight for better results. The percentage of undigested RNA can be estimated based on the signal seen at the origin during TLC analysis.

17. The solvent level in the reaction chamber must be below the spotted sample. The sample should not be in contact with the solvent until the solvent front has migrated up the plate as far as the origin.

Acknowledgments

We would like to thank all members of the Alfonzo Laboratory for their helpful discussions and comments. This work was supported in part by National Institutes of Health Grant GM084065 and Grant MCB0620707 from the National Science Foundation (to JDA).

References

1. Grosjean, H., Keith, G., and Droogmans, L. (2004) Detection and Quantification of Modified Nucleotides in RNA using Thin-Layer Chromatography. *Methods Mol. Biol.* *265*, 357–391.

2. Backus, J., and Smith, H. C. (1991) Apolipoprotein B mRNA Sequence 3' of the Editing Site are Necessary and Sufficient for Editing and Editosome Assembly. *NAR* *19*, 6781–6786.

3. Saiki, R. K., Scharf, S., Faloona, F., Mullis, K. B., Horn, G. T., Erlich, H. A., and Arnheim, N. (1985) Enzymatic Amplification of Beta-Globin Genomic Sequences and Restriction Site Analysis for Diagnosis of Sickle Cell Anemia. *Science.* *230*, 1350–1354.

4. Kotewicz, M. L., D'Alessio, J. M., Driftmier, K. M., Blodgett, K. P., and Gerard, G. F. (1985) Cloning and Overexpression of Moloney Murine Leukemia Virus Reverse

Transcriptase in Escherichia Coli. *Gene. 35,* 249–258.

5. Shuman, S. (1994) Novel Approach to Molecular Cloning and Polynucleotide Synthesis using Vaccinia DNA Topoisomerase. *J. Biol. Chem. 269,* 32678–32684.

6. Kapushoc, S. T., Alfonzo, J. D., and Simpson, L. (2002) Differential Localization of Nuclear-Encoded tRNAs between the Cytosol and Mitochondrion in Leishmania Tarentolae. *RNA. 8,* 57–68.

7. Alfonzo, J. D., Blanc, V., Estevez, A. M., Rubio, M. A., and Simpson, L. (1999) C to U Editing of the Anticodon of Imported Mitochondrial tRNA(Trp) Allows Decoding of the UGA Stop Codon in Leishmania Tarentolae. *EMBO J. 18,* 7056–7062.

8. Morl, M., Dorner, M., and Paabo, S. (1994) Direct Purification of tRNAs using Oligo-nucleotides Coupled to Magnetic Beads. in *Advances in Biomagnetic Separation* (E. H. Uhlen, and O. Olsvik, Eds.) pp 107–111, Easton Publishers, Natick, MA.

9. Chomczynski, P., and Sacchi, N. (1987) Single-Step Method of RNA Isolation by Acid Guanidinium Thiocyanate-Phenol-Chloroform Extraction. *Anal. Biochem. 162,* 156–159.

Chapter 14

Post-transcriptional Modification of RNAs by Artificial Box H/ACA and Box C/D RNPs

Chao Huang, John Karijolich, and Yi-Tao Yu

Abstract

RNA-guided RNA 2′-O-methylation and pseudouridylation are naturally occurring processes, in which guide RNAs specifically direct modifications to rRNAs or spliceosomal snRNAs in the nucleus of eukaryotic cells. Modifications can profoundly alter the properties of an RNA, thus influencing the contributions of the RNA to the cellular process in which it participates. Recently, it has been shown that, by expressing artificial guide RNAs (derived from naturally occurring guide RNAs), modifications can also be specifically introduced into other RNAs, thus offering an opportunity to study RNAs in vivo. Here, we present strategies for constructing guide RNAs and manipulating RNA modifications in the nucleus.

Key words: snoRNA, 2′-O-methylation, Pseudouridylation, snRNA, rRNA, mRNA, Telomerase RNA, Box C/D RNA, Box H/ACA RNA

1. Introduction

Present within eukaryotic cells is a large and diverse pool of metabolically stable noncoding RNAs (ncRNAs) (1). Originally identified in eukaryotic cells, small nucleolar RNAs (snoRNAs) and small Cajal body-specific RNAs (scaRNAs) are two such groups of ncRNAs, which primarily localize to subnuclear compartments referred to as the nucleolus and Cajal bodies, respectively (1–3). However, these two groups of RNA are not limited to eukaryotes; in fact, they are also found in abundance in archaea, single-celled organisms lacking nuclei. To date, several hundred of these RNAs have been described in various organisms ranging

Ruslan Aphasizhev (ed.), *RNA and DNA Editing: Methods and Protocols*, Methods in Molecular Biology, vol. 718,
DOI 10.1007/978-1-61779-018-8_14, © Springer Science+Business Media, LLC 2011

from archaea to mammals, making these RNAs one of the most abundant groups of ncRNAs (4–17).

Sno/sca RNAs can be further divided into two separate classes of RNAs, referred to, respectively, as Box C/D and Box H/ACA RNAs (Fig. 1) (18). Both Box C/D and Box H/ACA

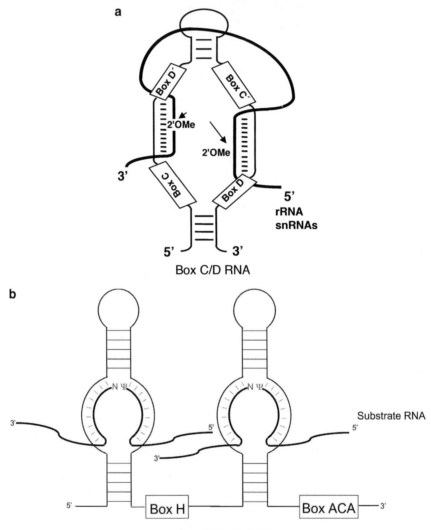

Fig. 1. (a) Schematic structure of a Box C/D RNA. The 5′ and 3′ end of the RNA adopts a terminal stem structure. The RNA has four conserved regions, from 5′ to 3′ termed Box C, Box D′, Box C′, and Box D. Between Box C and Box D′, and between Box C′ and Box D are guide sequences that are capable of basepairing with the complementary sequences within the substrate RNAs. The nucleotide which is basepaired to the fifth nucleotide upstream of Box D and/or Box D′ is the target for 2′-O-methylation. The thick black line represents the substrate RNA (including rRNA and snRNA). (b) Schematic representation of a Box H/ACA RNA. Box H/ACA RNAs adopt a hairpin-hinge-haipin-tail structure. The Box H is located within the hinge region, while the Box ACA is typically located three nucleotides upstream of the 3′ end. The internal loop is capable of basepairing with complementary sequences within the substrate RNA. The uridine residue targeted for pseudouridylation, as well the adjacent downstream nucleotide, are positioned at the base of the upper stem approximately 14–16 nucleotides upstream of either Box H or Box ACA and are left unpaired so as to remain accessible for isomerization.

RNAs exist in the cell as ribonucleoproteins (RNPs). The RNPs in each case consist of one sno/scaRNA and four class-specific core proteins. In the case of Box C/D RNPs, the protein components are Nop1p, Nop56p, Nop58p, and Snu13p (19–26), while for Box H/ACA RNPs the protein components are Cbf5p (dyskerin in humans), Gar1p, Nhp2p, and Nop10p (27–33). While some Box C/D and Box H/ACA RNPs are involved in the nucleolytic processing of rRNA, the majority function as modifying enzymes catalyzing, respectively, the site-specific post-transcriptional 2′-O-methylation and pseudouridylation of cellular RNAs (Fig. 2) (34–38). Nop1p and Cbf5p are the catalytic components of their respective RNPs (39, 40). Site-specificity is dictated by complementary basepairing interactions between the RNA component of the RNP and the substrate RNA. Given the strong evidence for RNP trafficking among subnuclear compartments, perhaps including the nucleoplasm, we believe that Box C/D and Box H/ACA RNPs are not confined within the nucleolus or Cajal bodies in eukaryotic cells. Consequently, Box C/D and Box H/ACA RNPs may be able to direct modifications to other cellular RNAs residing in various subnuclear compartments. Thus, by constructing novel Box C/D or Box H/ACA guide RNAs, we should be able to introduce 2′-O-methylation and pseudouridylation site-specifically into any RNAs in the nucleus.

Post-transcriptional modification of ribonucleotides represents a way to increase the chemical diversity of RNA molecules beyond the four canonical bases (adenine, guanine, cytosine, and

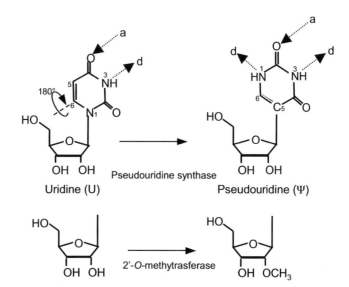

Fig. 2. The conversion of uridine to pseudouridine (*upper diagram*) and 2′-OH ribose to 2′-*O*-methylated ribose (*lower diagram*).

uracil) (41). While 2′-O-methylation is a sugar-ring modification, pseudouridylation is a uridine-specific modification. It is well established that both modifications alter the chemical properties of the nucleotide. These distinct chemical properties have the potential to impact numerous aspects of the RNA, including structure, thermal stability, and interactions (41). In each case, the effect of the modifications on the RNA depends on the structural context and can extend beyond the site of modification. In this regard, conformation stabilization appears to be an inherent property of pseudouridine, and is mediated by both an increase in base stacking and the ability to coordinate a water molecule through the extra hydrogen bond present (41–43). Furthermore, pseudouridine is slightly more polar than uridine (41). Similarly, data also suggest that 2′-O-methylation is capable of stabilizing RNA conformations. For instance, 2′-O-methylation blocks sugar edge interactions through the alteration of the hydration sphere around the oxygen (44–46). In addition, 2′-O-methylation inhibits the ability of the ribose to hydrogen bond with bases (41). Furthermore, it also plays a role in protecting the RNA from hydrolysis by alkaline substances and nucleases.

It is anticipated that, by introducing modified nucleotides into an RNA chain, we will be able to manipulate the properties of the RNA, thus offering an opportunity to study the structure of the RNA and its functions and mechanisms of action in the process in which it participates. Through the construction and expression of artificial guide RNAs, we have directed modifications to various cellular RNAs and have manipulated their ability to function as well as participate in various processes. For instance, the expression of an artificial C/D RNA capable of directing 2′-O-methylation to the branch point adenosine of *ACT1* pre-mRNA resulted in the modification of the residue and the subsequent inhibition of pre-mRNA splicing (47). Thus far, we have successfully targeted messenger RNAs (mRNA) ((47), Karijolich and Yu, unpublished data), spliceosomal small nuclear RNAs (snRNA) (Stephenson and Yu, Wu and Yu, unpublished data), ribosomal RNAs (rRNA) (Fig. 3), and telomerase RNA (Fig. 4, Huang and Yu, unpublished data). Here we present detailed strategies for both the construction and expression of a Box C/D RNA targeting rRNA and the mapping of novel 2′-O-methylation sites.

2. Materials

2.1. Construction of an snR52 Box C/D snoRNA Expression Cassette

1. Oligodeoxynucleotide primers: snR52-F1 5′-ACGTCGACA TAAATGATCT ACTATGATGAATGACATTATGCGCGC CTGCTTCTGATACAAAATCGAAAGATTTTAG GATTAGAA-3′, snR52-R1 5′-ATCTGCAGAAAAAATA AA

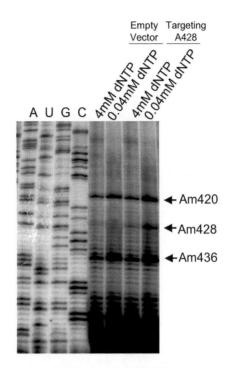

Fig. 3. Detection of 2'-O-methylation in yeast 18S rRNA by Primer Extension. Total RNA extracted from yeast harboring a plasmid targeting nucleotide A428 of 18S rRNA was assayed for novel 2'-O-methylation by primer extension. The Am420 and Am436 are naturally occurring 2'-O-methylation sites which are introduced by endogenous snR52 and endogenous snR87, respectively. A sequencing reaction was run in parallel to verify the modified nucleotide's identity.

TTTCAGAAGCAGGCGCGCATAAGTTTTTCTAA TCCTAAAATC-3' (IDT) (see Note 1).

2. 10× *Taq* DNA polymerase buffer (Fermentas).

3. 10 mM dNTPs (Fermentas).

4. *Pst*I restriction endonuclease (Fermentas).

5. *Sal*I restriction endonuclease (Fermentas).

6. 10× Orange restriction digestion buffer (Fermentas).

7. *Taq* DNA polymerase (5 U/μL) (Fermentas).

8. T4 DNA ligase (1 U/μL) (Fermentas).

9. 5× T4 ligase buffer (Fermentas).

10. DH5α competent cells (Stratagene).

11. LB liquid medium: 10 g NaCl, 10 g peptone, 5 g yeast extract, fill to 1 L with ddH$_2$O and autoclave.

12. LB-ampicillin solid medium; 20 g Agar, 10 g NaCl, 10 g peptone, 5 g yeast extract, fill to 1 L with ddH$_2$O and autoclave. Allow to cool and add 1 mL of 100 mg/mL ampicillin before pouring plates.

13. pSEC: snoRNA expression cassette (Fig. 5).

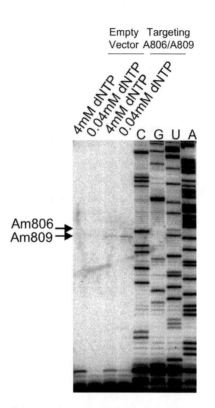

Fig. 4. Detection of 2′-0-methylation in yeast telomerase RNA by primer extension. Total RNA extracted from yeast harboring a plasmid targeting nucleotide A806 and U809 of TLC1 RNA was assayed for novel 2′-0-methylation by primer extension. A sequencing reaction was run in parallel to verify the modified nucleotide's identity. The two bands designated by *arrows* are A806 and U809 according to the sequence ladder.

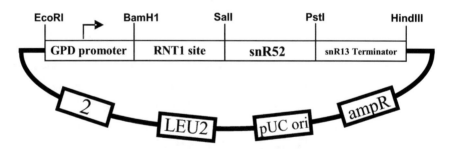

Fig. 5. Schematic representation of pSEC (containing artificial snR52). The artificial snR52 gene is flanked by two snoRNA processing elements (an RNT1 cleavage site and an snR13 terminator), and is expressed under the control of the GPD promoter. The RNT1 cleavage site represents the 65 nt long sequence in the 3′ ETS region of rRNA that is recognized and cut by the endonuclease Rnt1p (5′-TTT TTA TTT CTT TCT AAG TGG GTA CTG GCA GGA GTC GGG GCC TAG TTT AGA GAG AAG TAG ACT CA-3′); the snR13 terminator represents the 55 nt long sequence downstream of snR13 which is responsible for transcription termination (5′-AGT AAT CCT TCT TAC ATT GTA TCG TAG CGC TGC ATA TAT AAT GCG TAA AAT TTT C-3′). The restriction sites between each element are listed above every junction. pSEC is an *E. coli*-yeast shuttle plasmid. *2* Designates an origin of replication, and LEU2 is the auxotroph selective marker in yeast; pUC ori is the replication origin, and ampR is an ampicillin resistant marker in *E. coli*.

2.2. Transformation of Saccharomyces cerevisiae with snR52 Box C/D snoRNA Expression Cassette

1. One-Step-Transformation buffer: 100 mM lithium acetate, 50% (w/v) PEG-3350 solution.

2. YPD liquid medium: 10 g yeast extract, 20 g peptone, and 20 g dextrose, fill with ddH$_2$O to 1 L and autoclave.

3. SD-LEU liquid medium: 7.5 g Synthetic Leucine Drop Out Powder from Table 1, 20 g dextrose, fill to 1 L with ddH$_2$O and autoclave.

4. SD-LEU solid medium: 7.5 g Synthetic Leucine Drop Out Powder from Table 1, 20 g Agar, 20 g dextrose, fill to 1 L with ddH$_2$O and autoclave.

2.3. Total RNA Extraction from Saccharomyces cerevisiae

1. Trizol reagent (Invitrogen).

2. 0.5 mm acid washed glass beads (BioSpec).

3. Chloroform.

4. 10 mg/mL glycogen (Sigma).

5. 3 M Sodium acetate, pH 5.0.

6. Isopropanol.

2.4. Labeling and Purification of snR52-PXT Primer

1. T4 Polynucleotide Kinase (10 U/μL) (Fermentas).

2. snR52-PXT oligonucleotide 5′-GTTATTTATTGTCACTAC CTCCCTG-3′ (IDT) (see Note 2).

Table 1
Synthetic leucine drop out powder

Mixed powder	Amount for 15 L
Yeast nitrogen base	25.1 g
Ammonium sulfate	75.4 g
Isoleucine	450 mg
Valine	2.25 g
Adenine	300 mg
Arginine	300 mg
Histidine	300 mg
Leucine (dropped)	0 mg
Lysine	450 mg
Methionine	300 mg
Phenylalanine	750 mg
Tryptophan	300 mg
Tyrosine	450 mg
Uracil	300 mg

3. 10× T4 Polynucleotide Kinase Buffer A (Fermentas).

4. G50 Buffer: 20 mM Tris–HCl, 300 mM Sodium Acetate, 2 mM EDTA, 0.2% SDS, pH 7.5.

5. [γ-^{32}P] ATP (adenosine-5′-triphosphate. 6,000 Ci/mmol).

6. PCA: (phenol/chloroform/isoamyl alcohol = 25/24/1 [v/v/v]) saturated with 20 mM Tris–HCl, pH 8.0.

7. 40% Acrylamide:Bis (19:1).

8. 10% Ammonium persulfate (APS).

9. TEMED.

10. 5× TBE buffer: 445 mM Tris–HCl, 445 mM boric acid, 16 mM EDTA.

11. 2× Loading dye: 90% deionized formamide, 10 mM EDTA, 0.1% (w/v) bromophenol blue, 0.1% (w/v) xylene cyanol FF.

2.5. Detection of 2′-O-Methylation by Primer Extension

1. G50 Buffer: 20 mM Tris–HCl, 300 mM sodium acetate, 2 mM EDTA, 0.2% SDS, pH 7.5.

2. 2× annealing buffer: 500 mM KCl, 20 mM Tris–HCl, pH 8.3.

3. 2.5× dNTP 4 mM (High): 1 mM dATP, 1 mM dTTP, 1 mM dGTP, 1 mM dCTP, 8 mM DTT, 16 mM $MgCl_2$, 24 mM Tris–HCl, pH 8.3.

4. 2.5× dNTP 0.04 mM (Low): 0.01 mM dATP, 0.01 mM dTTP, 0.01 mM dGTP, 0.01 mM dCTP, 8 mM DTT, 16 mM $MgCl_2$, 24 mM Tris–HCl, pH 8.3.

5. Avian myeloblastosis virus (AMV) reverse transcriptase (10 U/μL) (Promega).

6. PCA: (phenol/chloroform/isoamyl alcohol = 25/24/1 [v/v/v]) saturated with 20 mM Tris–HCl, pH 8.0.

7. Ethanol.

8. 70% ethanol.

9. 2× Loading dye: 90% deionized formamide, 10 mM EDTA, 0.1% (w/v) bromophenol blue, 0.1% (w/v) xylene cyanol FF.

2.6. Sequencing Reactions

1. 5× sequencing buffer: 250 mM Tris–HCl, pH 9.0, 10 mM $MgCl_2$.

2. d/ddATP mixture: 350 μM ddATP, 80 μM dNTP.

3. d/ddCTP mixture: 200 μM ddCTP, 80 μM dNTP.

4. d/ddGTP mixture: 30 μM ddGTP, 80 μM dNTP.

5. d/ddTTP mixture: 600 μM ddTTP, 80 μM dNTP.

6. *Taq* DNA polymerase 5 U/μL (Promega).

7. p18SrRNA: plasmid-containing cDNA of yeast 18S rRNA corresponding to nucleotides 1–500.

3. Methods

3.1. Construction of an snR52 Box C/D snoRNA Expression Cassette

1. In a 0.2 mL PCR tube mix 5 μL of 10× *Taq* DNA polymerase buffer, 2 μL of 10 mM dNTPs, 2 μL of 10 μM snR52-F1 primer, 2 μL of 10 μM snR52-R1 primer, 1 μL (5 U/μL) of *Taq* DNA polymerase, and 38 μL of ddH$_2$O (see Fig. 6).

2. Perform PCR cycles as follows:

   Step 1: 95°C 2 min (1 cycle)

   Step 2: 95°C 30 s

           30°C 30 s

           72°C 30 s (repeat step 2 35 times)

   Step 3: 72°C 120 s (1 cycle)

           4°C (indefinitely)

3. Transfer PCR reaction to a 1.5-mL Eppendorf tube. Add 450 μL of G50 buffer and 500 μL of PCA. Vortex the mix for 30 s. Spin the mixture for 3 min at 13,000 rpm in a bench-top centrifuge.

4. Collect top aqueous phase and transfer to a clean 1.5-mL Eppendorf tube. Add 1 μL of 10 mg/mL glycogen and 1 mL of 100% ethanol.

5. Centrifuge sample at 13,000 rpm for 10 min in a bench-top centrifuge. Remove supernatant and allow the pellet to air dry for 5 min. Resuspend the pellet in 10 μL ddH$_2$O. Quantify the PCR product by U.V. spectroscopy.

6. In a 1.5-mL Eppendorf tube, add 5 μg of PCR products, 5 μL of 10× Orange digestion buffer, 2.5 μL of *Sal*I (10 U/μL) and 2.5 μL of *Pst*I (10 U/μL), and bring the total volume to 50 μL with ddH$_2$O. Incubate in a 37°C water bath overnight.

7. In a second 1.5 mL Eppendorf tube, add 1 μg of pSEC, 5 μL of 10× Orange digestion buffer, 2.5 μL of *Sal*I (10 U/μL) and

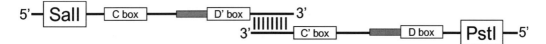

Fig. 6. snR52 PCR amplification strategy. Two long oligodeoxynucleotides were designed: one with the sense strand sequence of snR52 and the other with the antisense strand sequence of snR52. The guide sequences which are depicted as *gray boxes* were altered to basepair with the target sequences in the RNAs of interest. The two long oligodeoxynucle-otidedeoxy-oliognucleotides share a 15 nt complementary sequence at each 3′ end, which, when utilized in a PCR reaction, allows them to extend using each other as a template to generate an artificial snR52 mini-gene with specific restriction sites at each end (*Sal*I and *Pst*I in this case).

2.5 μL of PstI (10 U/μL), and bring the total volume to 50 μL with ddH$_2$O. Incubate in a 37°C water bath overnight.

8. To both tubes add 450 μL of G50 buffer, 500 μL of PCA, and vortex for 30 s. Centrifuge at 13,000 rpm for 5 min in a bench-top centrifuge.

9. Transfer the aqueous phase to new 1.5-mL Eppendorf tubes and add 1 mL of 100% ethanol, centrifuge at 13,000 rpm for 10 min in a bench-top centrifuge.

10. Remove supernatant and allow the pellet to air dry for 5 min. Dissolve both pellets in 10 μL of ddH$_2$O.

11. In a new 1.5-mL Eppendorf tube, add 5 μL of digested PCR product, 1 μL of digested plasmid, 2 μL of 5× T4 DNA ligase buffer, 2 μL of T4 DNA ligase (1 U/μL), and 10 μL of ddH$_2$O.

12. Incubate ligation mix in a 16°C water bath overnight.

13. Mix 10 μL ligation mix with 100 μL of DH5α competent cells, and put on ice for 10 min. Heat shock at 42°C for 45 s. Put on ice for 2 min.

14. Add 900 μL LB liquid medium and shake at 200 rpm for 45 min.

15. Centrifuge at 4,000 rpm for 5 min in a bench-top centrifuge.

16. Remove supernatant and resuspend cell pellet in 100 μL of LB liquid medium. Spread cells on LB-ampicillin solid medium plate.

17. Incubate in a 37°C incubator overnight to observe colonies.

3.2. Transformation of Saccharomyces cerevisiae with snR52 Box C/D snoRNA Expression Cassette

1. Inoculate single yeast colony into 5 mL YPD liquid medium and shake at 200 rpm overnight at 30°C.

2. Dilute cells in 5 mL fresh YPD medium to an OD$_{600nm}$ of 0.5.

3. When the optical density reaches 2 at OD$_{600nm}$, collect cells at 3,000 rpm using SH3000 rotor in a Sorvall RC-5C Plus centrifuge for 5 min at 4°C. Remove YPD medium completely (see Note 3).

4. Resuspend cell pellet in 200 μL of One-Step-Transformation Buffer to the concentration of 50 OD$_{600nm}$ cells/mL.

5. Aliquot 50 μL of Buffer/Cell mix to a new sterile 1.5 mL Eppendorf tube, add 1–5 μg of artificial snR52-containing pSEC and mix thoroughly.

6. Incubate mixture in 42°C water bath for 30 min.

7. Spin down cells at 3,000 rpm for 3 min at RT in a bench-top centrifuge, and remove supernatant.

8. Resuspend cells in 100 μL of ddH$_2$O and evenly spread on the surface of SD-LEU solid medium.

9. Incubate at 30°C for 2–3 days to observe colonies.

3.3. Total RNA Extraction from Saccharomyces cerevisiae

1. Inoculate single yeast colony (already transformed with artificial snR52-containing pSEC) in 20 mL of SD-LEU liquid medium. Shake at 200 rpm overnight at 30°C.

2. Collect the cells at 3,000 rpm for 5 min at 4°C using SH3000 rotor in a Sorvall RC-5C plus centrifuge.

3. Resuspend cell pellet in 1 mL of cold ddH$_2$O and transfer to a 2-mL screw-cap tube (see Note 4). Collect cells at 3,000 rpm for 5 min at 4°C using a chilled bench-top centrifuge.

4. Remove the ddH$_2$O and resuspend cells in 500 μL of Trizol reagent. Add 400 μL of acid washed glass beads.

5. Vigorously vortex the mixture for 1 min in a bench-top vortex. Place on ice for 1 min. Repeat 5 times.

6. Centrifuge at 13,000 rpm for 5 min at RT in a bench-top centrifuge. Transfer aqueous layer to a new 1.5-mL Eppendorf tube.

7. Add 100 μL of chloroform, vortex briefly and then centrifuge at 13,000 rpm for 5 min at RT in a bench-top centrifuge.

8. Transfer upper aqueous phase to a new 1.5 mL Eppendorf tube. Add 1 μL of 10 mg/mL glycogen and 500 μL of isopropanol and vortex briefly. Centrifuge at 13,000 rpm for 10 min at RT in a bench-top centrifuge.

9. Remove the supernatant and allow the pellet to air dry for 5 min. Resuspend the pellet in 20 μL of ddH$_2$O.

10. Measure RNA concentration by U.V./VIS spectroscopy. A good RNA preparation should give an A260/280 ratio of 1.8–2.0.

3.4. Labeling and Purification of snR52-PXT Primer

1. Mix 3 μL of 10 μM snR52-PXT primer with 1 μL of 10× PNK buffer A, 1 μL of [γ-^{32}P] ATP (10 μCi/μL), 1 μL of (10 U/μL) T4 PNK, and 4 μL of ddH$_2$O.

2. Incubate at 37°C for 30 min. Terminate reaction by adding 400 μL of G50 buffer and 400 μL of PCA.

3. Briefly vortex mixture. Centrifuge at 13,000 rpm for 5 min at RT in a bench-top centrifuge.

4. Transfer aqueous phase to a new 1.5-mL Eppendorf tube and add 1 μL of 10 mg/mL glycogen and 1 mL of cold ethanol.

5. Centrifuge at 13,000 rpm for 10 min in a bench-top centrifuge. Remove supernatant and allow the pellet to air dry for 5 min.

6. Dissolve pellet in 2 μL of ddH$_2$O and 2 μL of 2× Loading Dye. The sample is ready to be loaded.

7. Prepare an 8% polyacrylamide-8 M urea sequencing gel (16×30 cm, 0.4 mm spacer, 1 cm comb). For 60 mL of 8% polyacrylamide-8 M urea gel, dissolve 28.8 g urea in 12 mL 40% acrylamid/bisacrylamide (19:1) solution. Add 6 mL 5× TBE buffer and bring the final volume to 60 mL using ddH$_2$O. Add 360 μL 10% ammonium persulfate and 36 μL TEMED. Cast the gel immediately.

8. Pre-run the gel at 16 W of constant power for 30 min.

9. Denature the labeled primer at 95°C for 1 min and immediately load onto the gel.

10. Run the gel at 16 W of constant power. Run the gel until the bromophenol blue dye is approximately half way through the gel (see Note 5).

11. After removing the upper plate, wrap the gel in Saran Wrap plastic film, tape marker strips, and expose the gel to phosphorimager screen for 1 min (see Note 6).

12. Localize and cut the radiolabeled snR52-PXT primer form the gel. Place the gel slice into a 1.5-mL Eppendorf tube and add 500 μL of G50 buffer. Leave at RT overnight (see Note 7).

13. Add 500 μL PCA to G50 buffer/gel slice mix and vortex briefly. Centrifuge at 13,000 rpm for 5 min at RT in a bench-top centrifuge. Transfer the aqueous phase to a new 1.5-mL Eppendorf tube (see Note 8).

14. Precipitate snR52-PXT by adding 1 mL of cold ethanol and 1 μL of 10 mg/mL glycogen and vortex briefly. Centrifuge at 13,000 rpm for 10 min at RT in a bench-top centrifuge.

15. Remove the supernatant and add 1 mL of 70% ethanol. Centrifuge at 13,000 rpm for 10 min at RT in a bench-top centrifuge.

16. Remove the ethanol and allow the pellet to air dry for 5 min. Dissolve the pellet in 20 μL of ddH$_2$O and store at –20°C until use.

3.5. Detection of 2′-O-Methylation by Primer Extension

Perform two primer extension reactions for each RNA sample of interest. These two reactions are almost identical except for the amount of dNTP added. In the high dNTP concentration reaction, AMV can bypass sites of 2′-O-methylation site. However, reverse transcription is inhibited by the presence of a 2′-O-methylated residue when carried out at a low dNTP concentration. Specifically, the presence of a 2′-O-methylated residue will cause the AMV to stop one nucleotide before the modification site, which will appear as a premature stop when the reaction is resolved on a denaturing sequencing gel.

1. In a 1.5-mL Eppendorf tube mix 2 μL of 2× annealing buffer, 1 μL of labeled snR52-PXT, 5–20 μg of test RNA in a total of 4 μL volume, label as Mixture-High. Prepare a second tube exactly as above, but labeled as Mixture-Low (see Note 9).

2. Heat both Mixtures at 95°C for 1 min, and gradually chill at RT for 10 min.

3. In the meantime, prepare two different extension mixes. For Extension-High (High concentration dNTP), mix 16 μL of 2.5× dNTP 4 mM, 3 μL of ddH$_2$O, 1 μL of AMV (10 U). For Extension-Low (Low concentration dNTP), mix 16 μL of 2.5× dNTP 0.04 mM, 3 μL of ddH$_2$O, 1 μL of AMV (10 U) (see Note 10).

4. Add 4 μL of Extension-High to Mixture-High, and 4 μL of Extension-Low to Mixture-Low. Mix them well.

5. Incubate in a 42°C water bath for 30 min. Stop the reaction by adding 400 μL of G50 buffer. Add 400 μL PCA and vortex briefly. Centrifuge at 13,000 rpm for 5 min at RT in a bench-top centrifuge.

6. Transfer the upper aqueous phase into a new 1.5-mL Eppendorf tube and add 1 μL of 10 mg/mL glycogen, and 1 mL of cold ethanol.

7. Centrifuge at 13,000 rpm for 10 min at RT in a bench-top centrifuge.

8. Add 1 mL 70% cold ethanol and vortex briefly. Centrifuge at 13,000 rpm for 10 min at RT in a bench-top centrifuge.

9. Remove the ethanol and allow the pellet to air dry for 5 min. Dissolve the pellet in 3 μL of ddH$_2$O and 3 μL of 2× loading dye. Samples can be stored at –20°C for a couple of days (see Note 11).

3.6. Sequencing Reactions

To determine the position of 2'-O-methylation, a dideoxy-sequencing ladder should be run next to the primer-extension products from the modification mapping assay. In the sequencing reactions, a recombinant plasmid carrying the cDNA of the RNA of interest is used as a template (for both 18S rRNA and TLC1 RNA).

1. Mix 5 μL of 5× sequencing buffer, 1 μg of p18SrRNA, 2 μL of radiolabeled snR52-PXT, 1 μL of *Taq* DNA polymerase (5 U), and bring the final volume to 17 μL using ddH$_2$O. Label as Master Mix.

2. Label four PCR tubes as G, A, T, and C, respectively, and add 2 μL of the corresponding d/ddNTP mix to each PCR tube.

3. Distribute 4 μL of Master Mix to each PCR tube.

4. Perform PCR reaction as follows:

Step 1: 94°C 2 min (1 cycle)

Step 2: 94°C 30 s

    42°C 30 s

    72°C 30 s (repeat step 2 25 times)

Step 3: 4°C (indefinitely)

5. Add 6 μL of 2× loading dye to each tube and heat at 95°C for 2 min. Samples can be stored at –20°C for a couple of days.

3.7. Polyacrylamide Gel Electrophoresis

1. To analyze the primer-extension products and sequencing reaction products, prepare an 8% polyacrylamide-8 M urea sequencing gel (30 × 40 cm, 0.4 mm spacer, 0.5 cm comb). For an 80 mL of 8% polyacrylamide-8 M urea gel, dissolve 38.4 g urea in 16 mL of 40% acrylamide/bisacrylamide (19:1) solution. Add 8 mL 5× TBE buffer and bring the final volume up to 80 mL using ddH$_2$O. Add 480 μL of 10% ammonium persulfate and 48 μL of TEMED.

2. Pre-run the gel at 40 W of constant power for 30 min (see Note 12).

3. Denature primer extension and sequencing samples (see above) by heating at 95°C for 1 min immediately prior to loading.

4. Load 3 μL of primer extension reactions and 3 μL of each sequencing reaction on the gel. The remaining samples can be stored at –20°C.

5. Run the gel at 40 W of constant power until the bromophenol blue dye is approximately 4 cm from the bottom of the gel.

6. Separate the plates and transfer the gel onto a sheet of Whatman filter paper. Cover the gel side with Saran Wrap plastic film.

7. Dry gel in a BioRad Gel Dryer with the filter paper side facing the vacuum for 60 min at 90°C.

8. Expose dried gel to a phosphorimager screen overnight then visualize sequencing ladders and primer-extension products (see Note 13).

4. Notes

1. Artificial guide RNAs can be derived from any known Box C/D or Box H/ACA RNAs. Here we chose snR52 because it has two guide sequences available that can be altered to

target two unique sites of interest. We have also tested snR50 which has only one guide sequence available, and it also proved to be suitable for delivering artificial modification. The Box H/ACA RNA snR81 has been used for targeting mRNA and snRNA.

2. We usually use 22–25 nt long oligodeoxynucleotide primers which have a G or C at the 3′ end to ensure efficient extension in the reverse transcription reaction. If possible, design the primer such that it hybridizes approximately 40 nt downstream of the expected modification site. This will allow for the best resolution of the band representing the modified nucleotide.

3. We have experienced that if trace amounts of YPD medium are present in the transformation buffer the efficiency of transformation may be reduced. Washing the cell pellet with sterile ddH$_2$O before adding transformation buffer will increase the efficiency of transformation.

4. Washing yeast cells with sterile ddH$_2$O is necessary when using cells from saturation phase. Cell pellets can be stored at –70°C for a month without losing the integrity of total RNA.

5. When the bromophenol blue dye has ran approximately half way through the gel, the unreacted [γ-^{32}P] ATP and some radioactive free phosphate are still retained in the gel keeping the lower buffer chamber free of radioactivity and easing the cleaning of the gel box apparatus.

6. Depending on the intensity of [γ-^{32}P] ATP, it is recommended to titrate the exposure time from 30 s to 5 min.

7. Cutting the gel slice into a 5 × 5 mm square is ideal for downstream manipulation. Very small slices are difficult to be removed from the aqueous layer and can potentially hinder the precipitation of radiolabeled primers in ethanol.

8. The purpose of adding PCA is to keep the aqueous layer free of contamination by any trace amounts of protein, as well as keep the gel slice at the bottom of aqueous phase.

9. The minimum amount of RNA template needed for the reaction depends on the relative abundance of the RNA of interest. For example, 1 μg of total RNA is sufficient when determining the modification status of 18S ribosomal RNA. However, approximately 20 μg of total RNA is required when analyzing telomerase RNA.

10. Each 20 μL Extension mix is enough for four primer extension reactions (considering pippeting error). When dealing with a different number of reactions the mix can be scaled up or down.

11. It is recommended to load the primer-extension products onto the gel immediately after reactions. Although the remaining samples can be loaded and resolved on the gel later, we often experience that the immediate loading of the reactions results in the best looking gels.

12. When pre-running, one can load 3 µL of 2× loading dye into each well to ensure each lane is free of undesirable air bubbles.

13. In order to achieve best result, expose the gel for several days before scanning.

Acknowledgments

We would like to thank the members of the Yu Laboratory for discussion and inspiration. Our work was supported by grant GM62937 (to Yi-Tao Yu) from the National Institute of Health. J.K. was supported by a NIH Institutional Ruth L. Kirschstein National Research Service Award GM068411.

References

1. Matera, A. G., Terns, R. M., and Terns, M. P. (2007) Non-coding RNAs: lessons from the small nuclear and small nucleolar RNAs. *Nat. Rev. Mol. Cell. Biol.* **8**, 209–20.

2. Darzacq, X., Jady, B. E., Verheggen, C., Kiss, A. M., Bertrand, E., and Kiss, T. (2002) Cajal body-specific small nuclear RNAs: a novel class of 2′-O-methylation and pseudouridylation guide RNAs. *EMBO J* **21**, 2746–56.

3. Bachellerie, J. P., Cavaille, J., and Huttenhofer, A. (2002) The expanding snoRNA world. *Biochimie* **84**, 775–90.

4. Vitali, P., Royo, H., Seitz, H., Bachellerie, J. P., Huttenhofer, A., and Cavaille, J. (2003) Identification of 13 novel human modification guide RNAs. *Nucleic Acids Res.* **31**, 6543–51.

5. Huttenhofer, A., Kiefmann, M., Meier-Ewert, S., O'Brien, J., Lehrach, H., Bachellerie, J. P., et al. (2001) RNomics: an experimental approach that identifies 201 candidates for novel, small, non-messenger RNAs in mouse. *EMBO J* **20**, 2943–53.

6. Schattner, P., Barberan-Soler, S., and Lowe, T. M. (2006) A computational screen for mammalian pseudouridylation guide H/ACA RNAs. *RNA* **12**, 15–25.

7. Schattner, P., Decatur, W. A., Davis, C. A., Ares, M., Jr., Fournier, M. J., and Lowe, T. M. (2004) Genome-wide searching for pseudouridylation guide snoRNAs: analysis of the *Saccharomyces cerevisiae* genome. *Nucleic Acids Res.* **32**, 4281–96.

8. Yu, Y. T., Terns, R. M., and Terns M. P. (2005) Mechanisms and Functions of RNA-guided RNA Modification. In: Fine-Tuning of RNA Functions by Modification and Editing, Springer, Berlin, pp. 223–262.

9. Gu, A. D., Zhou, H., Yu, C. H., and Qu, L. H. (2005) A novel experimental approach for systematic identification of box H/ACA snoRNAs from eukaryotes. *Nucleic Acids Res.* **33**, e194.

10. Kiss, A. M., Jady, B. E., Bertrand, E., and Kiss, T. (2004) Human box H/ACA pseudouridylation guide RNA machinery. *Mol. Cell. Biol.* **24**, 5797–807.

11. Dunbar, D. A., Wormsley, S., Lowe, T. M., and Baserga, S. J. (2000) Fibrillarin-associated box C/D small nucleolar RNAs in *Trypanosoma brucei*. Sequence conservation and implications for 2′-O-ribose methylation of rRNA. *J Biol. Chem.* **275**, 14767–76.

12. Gaspin, C., Cavaille, J., Erauso, G., and Bachellerie, J. P. (2000) Archaeal homologs of eukaryotic methylation guide small nucleolar RNAs: lessons from the *Pyrococcus* genomes. *J Mol. Biol.* **297**, 895–906.

13. Omer, A. D., Lowe, T. M., Russell, A. G., Ebhardt, H., Eddy, S. R., and Dennis, P. P. (2000) Homologs of small nucleolar RNAs in Archaea. *Science* **288**, 517–22.

14. Qu, L. H., Meng, Q., Zhou, H., and Chen, Y. Q. (2001) Identification of 10 novel snoRNA gene clusters from *Arabidopsis thaliana*. *Nucleic Acids Res.* **29**, 1623–30.

15. Marker, C., Zemann, A., Terhorst, T., Kiefmann, M., Kastenmayer, J. P., Green, P., et al. (2002) Experimental RNomics: identification of 140 candidates for small non-messenger RNAs in the plant *Arabidopsis thaliana*. *Curr. Biol.* **12**, 2002–13.

16. Tang, T. H., Bachellerie, J. P., Rozhdestvensky, T., Bortolin, M. L., Huber, H., Drungowski, M., et al. (2002) Identification of 86 candidates for small non-messenger RNAs from the archaeon *Archaeoglobus fulgidus*. *Proc. Natl. Acad. Sci. USA* **99**, 7536–41.

17. Yuan, G., Klambt, C., Bachellerie, J. P., Brosius, J., and Huttenhofer, A. (2003) RNomics in Drosophila melanogaster: identification of 66 candidates for novel non-messenger RNAs. *Nucleic Acids Res.* **31**, 2495–507.

18. Balakin, A. G., Smith, L., and Fournier, M. J. (1996) The RNA world of the nucleolus: two major families of small RNAs defined by different box elements with related functions. *Cell* **86**, 823–34.

19. Lafontaine, D. L., and Tollervey, D. (2000) Synthesis and assembly of the box C+D small nucleolar RNPs. *Mol. Cell. Biol.* **20**, 2650–9.

20. Gautier, T., Berges, T., Tollervey, D., and Hurt, E. (1997) Nucleolar KKE/D repeat proteins Nop56p and Nop58p interact with Nop1p and are required for ribosome biogenesis. *Mol. Cell. Biol.* **17**, 7088–98.

21. Omer, A. D., Ziesche, S., Ebhardt, H., and Dennis, P. P. (2002) *In vitro* reconstitution and activity of a C/D box methylation guide ribonucleoprotein complex. *Proc. Natl. Acad. Sci. USA* **99**, 5289–94.

22. Ochs, R. L., Lischwe, M. A., Spohn, W. H., and Busch, H. (1985) Fibrillarin: a new protein of the nucleolus identified by autoimmune sera. *Biol. Cell* **54**, 123–33.

23. Galardi, S., Fatica, A., Bachi, A., Scaloni, A., Presutti, C., and Bozzoni, I. (2002) Purified box C/D snoRNPs are able to reproduce site-specific 2'-O-methylation of target RNA in vitro. *Mol. Cell. Biol.* **22**, 6663–8.

24. Watkins, N. J., Segault, V., Charpentier, B., Nottrott, S., Fabrizio, P., Bachi, A., et al. (2000) A common core RNP structure shared between the small nucleoar box C/D RNPs and the spliceosomal U4 snRNP. *Cell* **103**, 457–66.

25. Kuhn, J. F., Tran, E. J., and Maxwell, E. S. (2002) Archaeal ribosomal protein L7 is a functional homolog of the eukaryotic 15.5kD/Snu13p snoRNP core protein. *Nucleic Acids Res.* **30**, 931–41.

26. Lafontaine, D. L., and Tollervey, D. (1999) Nop58p is a common component of the box C+D snoRNPs that is required for snoRNA stability. *RNA* **5**, 455–67.

27. Henras, A., Henry, Y., Bousquet-Antonelli, C., Noaillac-Depeyre, J., Gelugne, J. P., and Caizergues-Ferrer, M. (1998) Nhp2p and Nop10p are essential for the function of H/ACA snoRNPs. *EMBO J* **17**, 7078–90.

28. Watkins, N. J., Gottschalk, A., Neubauer, G., Kastner, B., Fabrizio, P., Mann, M., et al. (1998) Cbf5p, a potential pseudouridine synthase, and Nhp2p, a putative RNA-binding protein, are present together with Gar1p in all H BOX/ACA-motif snoRNPs and constitute a common bipartite structure. *RNA* **4**, 1549–68.

29. Dragon, F., Pogacic, V., and Filipowicz, W. (2000) *In vitro* assembly of human H/ACA small nucleolar RNPs reveals unique features of U17 and telomerase RNAs. *Mol. Cell. Biol.* **20**, 3037–48.

30. Pogacic, V., Dragon, F., and Filipowicz, W. (2000) Human H/ACA small nucleolar RNPs and telomerase share evolutionarily conserved proteins NHP2 and NOP10. *Mol. Cell. Biol.* **20**, 9028–40.

31. Watanabe, Y., and Gray, M. W. (2000) Evolutionary appearance of genes encoding proteins associated with box H/ACA snoRNAs: cbf5p in *Euglena gracilis*, an early diverging eukaryote, and candidate Gar1p and Nop10p homologs in archaebacteria. *Nucleic Acids Res.* **28**, 2342–52.

32. Rozhdestvensky, T. S., Tang, T. H., Tchirkova, I. V., Brosius, J., Bachellerie, J. P., and Huttenhofer, A. (2003) Binding of L7Ae protein to the K-turn of archaeal snoRNAs: a shared RNA binding motif for C/D and H/ACA box snoRNAs in Archaea. *Nucleic Acids Res.* **31**, 869–77.

33. Wang, C., and Meier, U. T. (2004) Architecture and assembly of mammalian H/ACA small nucleolar and telomerase ribonucleoproteins. *EMBO J* **23**, 1857–67.

34. Kiss-Laszlo, Z., Henry, Y., Bachellerie, J. P., Caizergues-Ferrer, M., and Kiss, T. (1996) Site-specific ribose methylation of preribosomal RNA: a novel function for small nucleolar RNAs. *Cell* **85**, 1077–88.

35. Ganot, P., Bortolin, M. L., and Kiss, T. (1997) Site-specific pseudouridine formation in

preribosomal RNA is guided by small nucleolar RNAs. *Cell* **89,** 799–809.

36. Ni, J., Tien, A. L., and Fournier, M. J. (1997) Small nucleolar RNAs direct site-specific synthesis of pseudouridine in ribosomal RNA. *Cell* **89,** 565–73.

37. Bachellerie, J. P., Michot, B., Nicoloso, M., Balakin, A., Ni, J., and Fournier, M. J. (1995) Antisense snoRNAs: a family of nucleolar RNAs with long complementarities to rRNA. *Trends Biochem. Sci.* **20,** 261–4.

38. Cavaille, J., Nicoloso, M., and Bachellerie, J. P. (1996) Targeted ribose methylation of RNA *in vivo* directed by tailored antisense RNA guides. *Nature* **383,** 732–5.

39. Tollervey, D., Lehtonen, H., Jansen, R., Kern, H., and Hurt, E. C. (1993) Temperature-sensitive mutations demonstrate roles for yeast fibrillarin in pre-rRNA processing, pre-rRNA methylation, and ribosome assembly. *Cell* **72,** 443–57.

40. Zebarjadian, Y., King, T., Fournier, M. J., Clarke, L., and Carbon, J. (1999) Point mutations in yeast CBF5 can abolish *in vivo* pseudouridylation of rRNA. *Mol. Cell. Biol.* **19,** 7461–72.

41. Agris, P. F. (1996) The importance of being modified: roles of modified nucleosides and Mg^{2+} in RNA structure and function. *Prog. Nucleic Acid Res. Mol. Biol.* **53,** 79–129.

42. Arnez, J. G., and Steitz, T. A. (1994) Crystal structure of unmodified tRNA(Gln) complexed with glutaminyl-tRNA synthetase and ATP suggests a possible role for pseudo-uridines in stabilization of RNA structure. *Biochemistry* **33,** 7560–7.

43. Davis, D. R. (1995) Stabilization of RNA stacking by pseudouridine. *Nucleic Acids Res.* **23,** 5020–6.

44. Auffinger, P., and Westhof, E. (1997) Rules governing the orientation of the 2′-hydroxyl group in RNA. *J Mol. Biol.* **274,** 54–63.

45. Auffinger, P., and Westhof, E. (1998) Hydration of RNA base pairs. *J Biomol. Struct. Dyn.* **16,** 693–707.

46. Helm, M. (2006) Post-transcriptional nucleotide modification and alternative folding of RNA. *Nucleic Acids Res.* **34,** 721–33.

47. Zhao, X., and Yu, Y. T. (2008) Targeted pre-mRNA modification for gene silencing and regulation. *Nat. Methods* **5,** 95–100.

Chapter 15

Functional Analysis of Noncoding RNAs in Trypanosomes: RNA Walk, a Novel Approach to Study RNA–RNA Interactions Between Small RNA and Its Target

Chaim Wachtel and Shulamit Michaeli

Abstract

The recent discovery of thousands of small noncoding RNAs (ncRNAs), in many different organisms, has led to the need for methods to study their function. One way to help understand their function is to determine what other RNAs interact with the ncRNAs. We have developed a novel method to investigate the RNA–RNA interactions between a small RNA and its target that we termed "RNA walk." The method is based on UV-induced AMT cross-linking *in vivo* followed by affinity selection of the hybrid molecules and mapping the intermolecular adducts by RT-PCR. Domains carrying the cross-linked adducts are less efficiently amplified than domains that are not cross-linked. Real-time PCR is used to quantify the results. Further mapping of the interactions is performed by primer extension to determine the exact cross-linked adduct.

Key words: RNA walk, RNA–RNA interactions, Cross-linking, Affinity selection, Trypanosomes

1. Introduction

The recent identification of a large repertoire of small noncoding RNAs (ncRNAs) with unknown function in both eukaryotes and prokaryotes has led to the requirement for new methods to determine their possible function. ncRNAs exhibit a range of diverse functions, including ribosomal RNA maturation (small nucleolar RNAs), splicing of pre-mRNA (small nuclear RNAs), DNA replication (telomerase RNA), protein translocation (7SL RNA), and gene silencing (microRNAs or miRNAs). It has been suggested that ncRNAs are one of the major determinants of the complexity of an organism, complementing protein-based regulation (1, 2).

Ruslan Aphasizhev (ed.), *RNA and DNA Editing: Methods and Protocols*, Methods in Molecular Biology, vol. 718,
DOI 10.1007/978-1-61779-018-8_15, © Springer Science+Business Media, LLC 2011

The largest group of regulatory RNAs consists of miRNAs and siRNAs, with a length of only 21 nucleotides; these RNAs constitute the smallest class of ncRNAs described to date. miRNAs not only suppress translation via non-perfect base-pairing with the target 3′ UTR sequences of mRNAs, but can also elicit degradation of the target mRNA, and are central to a wide range of developmental and physiological processes (reviewed in (3)). Bioinformatic approaches have been used to suggest targets for many of the ncRNAs, but experimental methods are needed to validate such proposed interactions.

We have developed a novel method, termed "RNA Walk," to identify RNA–RNA interactions. While the method was developed in trypanosomes, it is a general method that can be used in any organism. In order to apply this method, one needs to determine the RNA molecule of interest that can be used as bait. In addition, it is desirable to have predictions for the small RNA target interaction, but it is not absolutely necessary. The approach, which is outlined schematically in Fig. 1, utilizes AMT-induced UV cross-linking, which covalently links two RNA molecules that are located in close proximity by intercalating between the double-helical region of the nucleic acids (4). The cells are incubated with AMT on ice and then irradiated at 365 nm. After *in vivo* cross-linking, total RNA is isolated and the hybrid cross-linked species are affinity selected, either using the small RNA or the known target as bait. The position of the interaction between the small RNA and the target is mapped by "walking" on the target RNA using RT-PCR and primers that amplify the proposed target, identifying domains that cannot be copied efficiently due to the presence of cross-linked adducts (see Fig. 1). We previously used this method to map four interactions between a special tRNA-like molecule (sRNA-85) that is part of the trypanosome signal recognition particle (SRP) complex (5, 6) and the ribosome (7). Since the method relies on AMT-induced UV cross-links there might be additional contact sites that the "RNA walk" fails to identify because the AMT cross-linking is limited to uridines which are in close proximity within the interaction domain, although cytidines were also shown to be cross-linked (4). The critical step of the method is the enrichment of the duplex by affinity selection. This avoids mapping of intramolecular cross-links and enriches for intermolecular cross-linking. This method is a generic method that we used to map the position of the interaction of sRNA-85 and its known target, the rRNA. However, this approach can also be used to determine the target using various techniques including the recently developed RNA sequencing (8).

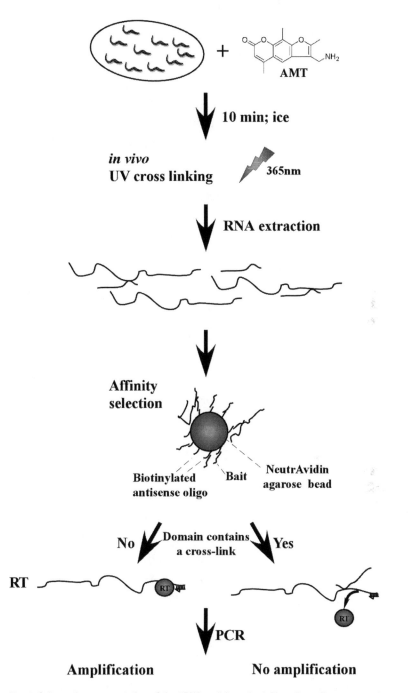

Fig. 1. Schematic representation of the "RNA walk" method. First, the cells are treated with AMT on ice for 10 min and subjected to UV irradiation at 365 nm on ice. Total RNA is extracted from the cells and the RNA of interest (the bait) is affinity selected, along with any cross-linked hybrids. cDNA is prepared from the affinity-selected RNA and amplified by PCR using primers that cover the entire target.

2. Materials

2.1. Cross-Linking and RNA Extraction

1. Dulbecco's phosphate buffered saline, 10×, modified, without calcium chloride and magnesium chloride, liquid, sterile-filtered, cell culture tested (Sigma). Dilute to 1× with Mili-Q grade water (see Note 1).

2. 4′-Aminomethyltrioxsalen hydrochloride (AMT) (Sigma). Prepare stock solution of 1 mg/ml in water and store at 4°C in the dark, do not freeze.

3. Proteinase K, store 10 mg/ml stock solution at −20°C and prepare 200 μg/ml solution in 1% SDS before use.

4. TRI Reagent® (Sigma). Store at 4°C.

2.2. Affinity Selection

1. 5× hybridization buffer: 100 mM HEPES (pH 8), 25 mM $MgCl_2$, 1,500 mM KCl, 0.05% NP-40, 5 mM dl-Dithiothreitol (DTT).

2. Biotinylated (at 3′ end) RNA or DNA oligonucleotide complimentary to the small RNA.

3. NeutrAvidin agarose resin (Pierce). Store at 4°C.

4. Blocking buffer: 200 μl WB100 (see below), 50 μl BSA (10 mg/ml stock solution), 40 μl tRNA (10 mg/ml stock solution), 10 μl glycogen (20 mg/ml, Roche), and 700 μl DEPC (diethyl pyrocarbonate)-treated water.

5. WB100 buffer: 20 mM HEPES (pH 8), 10 mM $MgCl_2$, 100 mM KCl, 0.01% NP-40, 1 mM DTT.

6. WB400 buffer: 20 mM HEPES (pH 8), 10 mM $MgCl_2$, 400 mM KCl, 0.01% NP-40, 1 mM DTT.

2.3. PCR

1. RevertAid™ First Strand cDNA synthesis kit (Fermentas), store at −20°C.

2. ReadyMix™ Taq PCR reaction mix with $MgCl_2$ (Sigma), store at −20°C.

3. Absolute Blue QPCR SYBR® green ROX mix (Thermo Scientific).

2.4. Primer Extension

1. [γ-^{32}P]-ATP, 3,000 Ci/mmol, 10 mCi/ml, store at 4 or −20°C as per suppliers' instructions.

2. T4 polynucleotide kinase, store at −20°C.

3. Expand reverse transcriptase, 100 mM DTT and 5× reaction buffer (Roche), store at −20°C.

4. 5 mM dNTPs, many companies sell 10 mM dNTP mixes that can be diluted 1:2 for the primer extension reaction.

5. RNase inhibitor, store at −20°C.

2.5. RNase H Cleavage and Splint Labeling

1. RNase H, store at –20°C.

2. Sequenase™ Version 2.0 DNA Sequencing Kit (USB), store at –20°C.

3. [α-^{32}P]-dCTP, 3,000 Ci/mmol 10 mCi/ml, store at 4 or –20°C as per suppliers instructions.

4. RNase H cleavage buffer (10×): 750 mM KCl, 500 mM Tris–HCl (pH 8.3), 30 mM MgCl$_2$, 100 mM DTT.

5. Splint labeling hybridization buffer: 50 mM Tris–HCl (pH 7.8), 10 mM MgCl$_2$, and 1 mM DTT.

6. Phenol/chloroform (1:1) store at 4°C.

7. 100 and 75% ethanol.

8. Proteinase K buffer: 50 mM Tris–HCl (pH 7.5); 10 mM EDTA; 100 mM NaCl; 1% SDS.

2.6. Denaturing Poly-Acrylamide Gel Electrophoresis

1. 10× TBE: 108 g Tris base (final concentration 90 mM), 55 g boric acid (final concentration 90 mM), and 40 ml 0.5 M EDTA (pH 8.0) (final concentration 2 mM) in total volume of 1 l. Store at room temperature.

2. 6% poly-acrylamide gel mix: 100 ml 30% 19:1 acrylamide/bis solution, 210 g urea (7 M final concentration), and 50 ml 10×TBE in a final volume of 500 ml. Filter sterilize and store at 4°C.

3. 15% poly-acrylamide gel mix: 250 ml 30% 19:1 acrylamide/bis solution, 210 g urea (7 M final concentration), and 50 ml 10× TBE in a final volume of 500 ml. Filter sterilize and store at 4°C.

4. N,N,N,N'-Tetramethyl-ethylenediamine (TEMED); store at 4°C and ammonium persulfate (10% solution in water), freeze in small aliquots and store at –20°C.

5. RNA loading buffer: 95% formamide, 0.05% bromophenol blue, 0.05% xylene cyanol, and 20 mM EDTA.

6. Elution buffer: 0.3 M sodium acetate, 10 mM ethylenediaminetetraacetic acid (EDTA).

3. Methods

"RNA walk" is a novel method that combines *in vivo* AMT-induced UV cross-linking, affinity selection, and RT-PCR or preferentially quantitative real-time PCR. The method takes advantage of the fact that AMT cross-linking, which links two RNA molecules that are located in close proximity, blocks reverse transcription. Therefore, domains carrying the cross-linked adducts fail to efficiently amplify by PCR compared with non-cross-linked domains (see Fig. 1). The cross-linking is performed at 365 nm and can be reversed when irradiated at 254 nm. The affinity selection step ensures that domains that interact with the small RNA of interest will not be amplified when compared to domains that do not contain cross-linked adducts. Further validation of the "RNA walk" can be performed using a number of different methods, two of which, primer extension mapping and RNase H cleavage coupled with splint labeling, are detailed below.

3.1. In Vivo *Cross-Linking of Leptomonas collosoma Cells*

1. Grow cells at 28°C to a concentration of 5×10^7 cells/ml in 500 ml of BHI (see Note 2).

2. Centrifuge cells at $3,000 \times g$, 4°C for 10 min.

3. Wash twice in 1× PBS.

4. Resuspend the cell pellet in 8 ml 1× PBS and split to three tubes, as follows: tube #1, 3.2 ml, tube #2, 3.2 ml, and tube #3, 1.6 ml. We find that there are two controls that are required at this stage: no irradiation and irradiation in the absence of AMT (see Note 3).

5. Add 0.8 ml AMT to tube #1 (final concentration of 0.2 mg/ml), and add 1× PBS to tubes #2 and #3 to a final volume of 4 ml in each tube (0.8 ml for tube #2 and 2.4 ml for tube #3). This step should be done in the dark as the AMT is light sensitive. It is advised to wrap tube #1 in aluminum foil in order to prevent light from damaging the AMT.

6. Incubate the tubes on ice for 10 min.

7. Transfer the cells from tubes #1 and #2 to separate premarked 35×10 cell culture dishes. Use a number of dishes so that the cell spread is spread on the plate in a single layer. Place the dishes on ice in such a manner that no ice can enter the dish and contaminate the sample. In order to achieve this, we place the dish on a metal block in ice (see Fig. 2). The ice is placed on a rotator to ensure that all cells are irradiated. Cross-link

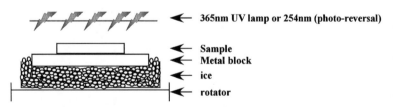

Fig. 2. Schematic representation of *in vivo* UV irradiation set-up. In order to limit the amount of RNA degradation during irradiation the cells are kept on ice throughout the process. The cells are placed in a number of 35×10 plates, the amount should be sufficient that the cell suspension covers most of the plate. The plate itself is placed on a metal block that is sitting in a box of ice. The ice box with the samples is placed on a rotator to allow for rotation of the cells during irradiation, to help ensure that all cells are exposed to the UV lamp. The UV lamp is placed on top of the ice box, and the whole apparatus rotates for the duration of the irradiation.

the tubes #1 and #2 for 60 min (for *Trypanosoma brucei* cross-link for 30 min) on ice using a UV lamp at 365 nm with an intensity of 10 mW/cm² (see Note 4). While irradiating the cells from tubes #1 and #2 leave tube #3 on ice.

8. Wash the cells once with 1× PBS.

9. Deproteinize the cells by digestion with proteinase K (200 μg/ml) for 60 min at 37°C.

10. Prepare RNA using TRI Reagent® (Sigma, catalog #T9424) according to the manufacturer's instructions.

3.2. Affinity Selection

1. Take 500 μg of each RNA sample in 1× hybridization buffer and heat at 80°C for 2 min, and immediately cool on ice.

2. Add 8 nmol of anti-sense biotinylated RNA (or DNA) oligonucleotide to each tube, and mix well by vortex.

3. Incubate at room temperature over-night.

4. Prepare beads as follows: wash 50 μl of NeutrAvidin agarose resin (50% slurry) for each RNA sample in 500 μl WB100 two times. Block beads in 1 ml of blocking buffer, rotating at 4°C for at least 2 h (see Note 5).

5. Wash beads one time in blocking buffer (without BSA, tRNA or glycogen).

6. Add hybridized RNA samples, from step 2, to beads and bind to beads in rotator at 4°C for at least 2 h (see Note 6). In our case, 500 μg of total RNA contained 0.2 pmol of sRNA-85, of which approximately 20% can be recovered by affinity selection with 50 μl of beads.

7. Wash beads five times with 500 μl of WB400 buffer for 30–60 s per wash.

8. Elute the RNA from the beads by adding 1 ml of TRI Reagent® and continue isolating the RNA according to the manufacturer's instructions. A second and third round of affinity selection can be performed. However, while we do see some further enrichment of the small RNA (sRNA-85), we lost ~90% of the contaminating rRNA after one round of selection and lost 100% of the nonrelated sno-2 RNA (9). Therefore, it may not be necessary to perform more than one round of affinity selection, although this should be tested for each system examined.

9. Resuspend the eluted RNA in 50 μl of water or other buffer of choice.

3.3. RT-PCR

1. Take one-tenth of the volume of each of the affinity-selected RNAs and prepare cDNA using random hexamer or a gene-specific primer and the RevertAid™ First Strand cDNA synthesis kit following the manufacturer's instructions (see Note 7).

2. Perform a standard PCR reaction on the potential target of your RNA. In our study we divide the ribosomal RNAs into 12

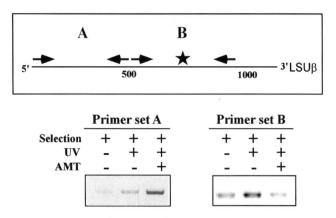

Fig. 3. "RNA walk" on the LSUβ subunit of rRNA. PCR analyses of cDNA covering the rRNA. RNA was prepared from irradiated cells and from control untreated cells as described in Subheading 3.1. The RNA (from 10^{10} cells) was subjected to affinity selection as described in Subheading 3.2. The affinity-selected product was subjected to reverse transcription using random primers. The cDNA was amplified by PCR using primers dividing the LSUβ subunit of rRNA into 2,500 nt domains as depicted in the upper panel. The positions of the PCR domains are marked by arrows on the LSUβ subunit of rRNA (indicated). A star marks the domain carrying the cross-linked adducts. The PCR products were separated on a 1.5% agarose gel and detected by ethidium bromide staining (*lower panel*).

domains of approximately 500 bases each and amplified each of them looking for interactions with sRNA-85 (see Fig. 3 for an example). The number of cycles for each PCR reaction must be optimized to ensure exponential amplification. The concentration of cDNA used in the reaction must also be optimized in order to observe the differences in amplification ability between the two different types of domains. Regions that contain cross-linked adducts will be amplified poorly as compared to regions that do not contain adducts when the proper amount of cDNA is used in the reaction (see Note 8).

3. Once domains that interact with the small RNA have been identified, the region of interaction can be further narrowed down using more internal primer sets for PCR. For example, we initially looked at domains 500 nucleotides long and then narrowed them down by PCR to 100 nucleotide fragments. Then these regions are used for fine mapping by primer extension (see Subheading 3.5).

3.4. Quantitative Real-Time PCR

1. In order to quantify the reduction in cross-linked regions as compared to non-cross-linked regions, real-time PCR should be performed.

2. Perform real-time PCR using SYBR green from Thermo Scientific, using the primers previously used for RT-PCR (see Note 9). Due to the sensitivity of real-time PCR it is necessary to dilute the cDNAs more drastically than the reactions described in Subheading 3.3. For example, while we diluted

our cDNA 1:500 for analysis by PCR, for real-time PCR we had to dilute the cDNA 1:100,000.

3. In order to obtain absolute amounts of PCR products produced, we first established a calibration curve for each set of primers examined. First, cDNA from total RNA was amplified by PCR. This PCR product was diluted 1:1,000, and used as a template for a second round of PCR using the same primers. An aliquot of the PCR reaction is analyzed on an agarose gel to ensure that a single PCR product is produced. The remainder of the PCR fragment is purified by the QIAquick PCR purification kit (Qiagen) and its concentration is determined by NanoDrop ND-1000 spectrophotometer (Thermo). To generate a calibration curve, the purified PCR fragment is diluted to different concentrations (we used five different concentrations) and real-time PCR reactions are performed. The software supplied with the real-time PCR machine is then used to generate the calibration curve.

3.5. Primer Extension

1. In order to identify the exact nucleotide adduct between the small RNA and the target, primer extension is performed. First, label an anti-sense primer, located within the 3′ end of the interacting domain (see Note 10), with $[\gamma\text{-}^{32}P]$-ATP as follows: take 100–200 pmol oligonucleotide, 2 µl $[\gamma\text{-}^{32}P]$-ATP, 2 µl 10× buffer (supplied with the enzyme), 1 µl polynucleotide kinase (10 units, NEB), and water to a final volume of 20 µl. Incubate the reaction at 37°C for 1 h.

2. Gel purify the labeled primer as follows: first, prepare a 15×17×0.04 cm 15% denaturing gel for vertical gel electrophoresis. For a gel of this size take 10 ml of gel mix, add 10 µl TEMED and 100 µl 10% APS (ammonium per sulfate), and immediately pour the gel. After the gel has polymerized (30–60 min), pre-run the gel in 1× TBE for 15–30 min at 200 V. Add 5–10 µl of RNA loading buffer to the primer, and heat the sample for 2 min at 90°C. Load the sample on the gel and run till the bromophenol blue is near the bottom. Expose the gel to film for 30–60 s and cut out the labeled primer from the gel. Elute the gel slice in 400 µl of elution buffer at 37°C for at least 2 h (see Note 11). Transfer the liquid to a new tube and add 1 ml 100% ethanol. Precipitate at –80°C for at least 15 min, wash with 75% ethanol, and resuspend in ~50 µl water.

3. Perform primer extension on 1/10 of the affinity-selected RNAs as well as on 5 µg of total RNA (both +UV and –UV) as follows: to each RNA sample add 5 µl 5× reaction buffer (supplied with the enzyme), 2.5 µl DTT (100 mM, supplied with the enzyme), 3 µl dNTPs (5 mM), 50,000 cpm labeled oligonucleotide, and water to a final volume of 24 µl. Heat the samples at 60°C for 15 min. Cool samples on ice for 1 min

and add 1 μl of RT mix (1 μl Expand reverse transcriptase, 1 μl RNase inhibitor, and 8 μl water). Incubate the reaction at 42°C for 90 min. Precipitate the primer extension products by adding 62.5 μl 100% ethanol, 2.5 μl sodium acetate, and place at –80°C for at least 15 min. Centrifuge the samples in a micro-centrifuge at maximum speed (20,800×g) for 30 min at 4°C and resuspend in 12 μl RNA loading buffer.

4. Analyze the primer extension reaction by gel electrophoresis. Prepare a 35×20×0.04 cm 6% denaturing gel. For a gel of this size 30–40 ml of gel mix is required. Add 300 μl 10% APS and 24 μl TEMED, quickly pour the gel and let it polymerize for ~1 h. Pre-run the gel for 30 min at 400 V, meanwhile heat samples for 2 min at 90°C, and immediately place on ice. Load half the samples on the gel and run the gel at 1,000 V till the bromophenol blue is near the bottom of the gel. Run a DNA sequencing experiment in the lanes next to the primer extension samples in order to identify the cross-linked adduct (see Note 12). Transfer the gel to a piece of Whatman paper and dry in a gel drier. Expose the dried gel to either film or a phosphorimager screen (see Note 13).

3.6. RNase H Cleavage and Splint Labeling

1. Take 5–10 μg of total RNA and one-tenth of the affinity selected RNAs (~50 ng) in a tube. Add 200 pmol of each anti-sense primer (see Note 14), 2 μl 10× buffer (supplied with the enzyme) and water to a final volume of 19 μl. Incubate samples at 65°C for 10 min and place on ice. Add 1 unit of RNase H (we use a 1:5 dilution of the stock supplied by NEB) and incubate the reaction at 37°C for 60–90 min.

2. Add 200 μl proteinase K buffer, 1 μl proteinase K and 0.5 μl glycogen to each sample and incubate at 37°C for 15–30 min (see Note 15). Add 250 μl phenol:chloroform, vortex, centrifuge in a micro-centrifuge at maximum speed (20,800×g) for 5–10 min, and transfer the samples to new tubes. Precipitate the RNA by adding 500 μl ethanol and placing at –80°C for at least 15 min. Centrifuge the samples in a micro-centrifuge at maximum speed (20,800×g) for 30 min at 4°C and resuspend in 15 μl water.

3. Label the RNA by splint labeling as follows: take the RNA from step 2 and heat at 85°C for 2 min.

4. Add 2 μl splint labeling hybridization buffer, 100 pmol of splint labeling primer (see Note 16), vortex and quench on ice for 30 min. Add 50 μCi of [α-^{32}P]-dCTP (3,000 Ci/mmol) and 5 units of T7 DNA polymerase (Sequenase v. 2.0, USB), and incubate for 1 h at 37°C. Precipitate the samples as described in step 2 and resuspend in 12 μl RNA loading buffer. Run the samples on a 6% poly-acrylamide gel as described in Subheading 3.4, step 4.

4. Notes

1. Work with RNA requires RNase-free water. We use Mili-Q grade double-distilled water (DDW). From here on, when we refer to water we mean Mili-Q grade DDW.

2. The conditions of cell growth (amount, type of medium, and density of cells) should be adjusted depending upon the type of cells being used in the study. For *Leptomonas collosoma* we grow the cells in Brain Heart Infusion (BHI) with hemin (25 µg/l) and penicillin–streptomycin (4 ml/l). For *Trypanosoma brucei* we grow the cells in SDM-79 medium supplemented with 10% fetal calf serum (10). We typically use 500 µg of RNA per affinity selection reaction. The amount of cells grown for cross-linking should be adjusted depending on the type of cells used and the abundance of the RNA.

3. The amount of cells can be split differently, but we found that RNA degradation occurs due to the UV treatment and it is best to use more cells for the samples that undergo irradiation. The analysis of AMT cross-linking experiments must be tightly controlled as they are prone to artifacts. As mentioned above, UV irradiation causes RNA degradation which may interfere with the analysis of the experiment. Therefore, one important control is to irradiate cells in the absence of AMT in addition to examining cells that were not irradiated. To further show that any stops observed are due to the AMT treatment, it is necessary to reverse the cross-links by irradiating at 254 nm.

4. The length of exposure to UV light must be tested empirically. We tested different time points from 15 to 60 min and found that 60 min gave the best results for *Leptomonas collosoma* and 30 min for *Trypanosoma brucei*.

5. For convenience blocking of the beads can be done overnight.

6. We have hybridized 2–4 h without seeing any difference in the end result.

7. While we used this kit from Fermentas, other methods of performing RT can be used. We observed no difference when we used random hexamer or a gene-specific primer. However, we prefer to use random hexamer due to its versatility. Using random hexamer allows one to examine many potential interacting domains using the same cDNA preparation.

8. For each domain we tested at least four different cDNA concentrations: 1:10, 1:100, 1:500 and 1:1,000. We obtained the best results when we diluted the cDNA either 1:500 or 1:1,000. However, we were studying a very abundant RNA (rRNAs) and one should test a number of different dilutions

when setting up their experiments with less abundant RNAs. We typically found that 25 cycles was within the linear phase of the PCR reaction, but this must be tested for each gene that is used.

9. We used SYBR green from Thermo Scientific, which worked identically to the regular PCR we performed. However, one may use any real-time PCR system that works for them.

10. We typically use primers 15–18 nucleotides long. The rest of the section is based on using oligonucleotides of this length. The primer should be located no more than 100 nucleotides from the cross-linked site, as extensions longer than that can be difficult to analyze. Therefore, more than one primer may be required to determine the exact cross-link site.

11. Typically we recover 50–75% of the label after 2 h of elution when using a 0.04 cm thick gel. When a thicker gel is used longer incubation times are required (up to over-night).

12. In order to determine the cross-linked adduct it is necessary to perform a DNA sequencing reaction using the identical primer that is used for the primer extension reactions. This can be performed either using a cloned DNA fragment or a PCR product as a template. We used the Sequenase™ Version 2.0 DNA Sequencing Kit (USB, catalog #70770) and followed the manufacturer's instructions.

13. When analyzing the primer extension data one must be very careful as besides stops resulting from the AMT cross-linking between two molecules there can be stops from intramolecular cross-links. Besides AMT-induced stops one can also get stops if the RNA molecule is modified (methylation for example), as well as due to degradation of the RNA during the course of the experiment. Therefore, a number of controls similar to the ones used for the "RNA walk" must be performed when performing the experiments. These include performing primer extension on RNA that was not affinity selected, on RNA that was not UV cross-linked, and on RNA in which the AMT cross-linking is reversed by irradiation at 254 nm.

14. For each interacting domain that we identified by "RNA walk" we chose anti-sense primers that reduce the domain carrying the cross-link to less than 500 nucleotides. This is performed in order for the cross-linked species to be able to enter a poly-acrylamide gel, which can resolve RNA molecules that are less than 1,000 nucleotides long.

15. Proteinase K treatment is critical if one wants to label the cross-linked species using the splint-labeling method. If the RNase H is not sufficiently removed it may interfere with splint labeling.

16. Splint labeling is a technique in which a primer complementary to the RNA you want to label carrying a run of N nucleotides at the 3′ end of the oligonucleotide is used as a template to incorporate the complementary $[\alpha\text{-}^{32}\text{P}]$-dNTP at the 3′ end of the RNA using the enzyme T7 DNA polymerase (11). For example, we used an oligonucleotide with a run of nine G nucleotides at the 3′ end and labeled the RNA using $[\alpha\text{-}^{32}\text{P}]$-dCTP. It is required to know the exact end of the RNA you are labeling in order to use this technique to label all RNA molecules that contain the bait.

Acknowledgement

This research was supported by a grant from the Israel–US Binational Science Foundation (BSF), and by an International Research Scholar's Grant from the Howard Hughes Foundation to S.M. S.M. holds the David and Inez Myers Chair in RNA silencing of diseases.

References

1. Mattick, J. S., and Makunin, I. V. (2006) Non-coding RNA, *Hum Mol Genet* 15 *Spec No* 1, R17–R29.

2. Mercer, T. R., Dinger, M. E., and Mattick, J. S. (2009) Long non-coding RNAs: insights into functions, *Nat Rev Genet* 10, 155–159.

3. Chekulaeva, M., and Filipowicz, W. (2009) Mechanisms of miRNA-mediated post-transcriptional regulation in animal cells, *Curr Opin Cell Biol* 21, 452–460.

4. Cimino, G. D., Gamper, H. B., Isaacs, S. T., and Hearst, J. E. (1985) Psoralens as photo-active probes of nucleic acid structure and function: organic chemistry, photochemistry, and biochemistry, *Annu Rev Biochem* 54, 1151–1193.

5. Liu, L., Ben Shlomo, H., Xu, Y. X., Stern, M. Z., Goncharov, I., Zhang, Y., and Michaeli, S. (2003) The trypanosomatid signal recognition particle consists of two RNA molecules, a 7SL RNA homologue and a novel tRNA-like molecule, *J Biol Chem* 278, 18271–18280.

6. Beja, O., Ullu, E., and Michaeli, S. (1993) Identification of a tRNA-like molecule that copurifies with the 7SL RNA of *Trypanosoma brucei*, *Mol Biochem Parasitol* 57, 223–229.

7. Lustig, Y., Wachtel, C., Safro, M., Liu, L., and Michaeli, S. (2010) 'RNA walk' a novel approach to study RNA-RNA interactions between a small RNA and its target, *Nucleic Acids Res* 38, e5.

8. Wang, Z., Gerstein, M., and Snyder, M. (2009) RNA-Seq: a revolutionary tool for transcriptomics, *Nat Rev Genet* 10, 57–63.

9. Levitan, A., Xu, Y. X., Ben Dov, C., Ben Shlomo, H., Zhang, Y. F., and Michaeli, S. (1998) Characterization of a novel trypanosomatid small nucleolar RNA, *Nucleic Acids Res* 26, 1775–1783.

10. Brun, R., and Schonenberger, M. (1979) Cultivation and *in vitro* cloning or procyclic culture forms of *Trypanosoma brucei* in a semi-defined medium. Short communication, *Acta Trop* 36, 289–292.

11. Hausner, T. P., Giglio, L. M., and Weiner, A. M. (1990) Evidence for base-pairing between mammalian U2 and U6 small nuclear ribonucleoprotein particles, *Genes Dev* 4, 2146–2156.

Chapter 16

A Post-Labeling Approach for the Characterization and Quantification of RNA Modifications Based on Site-Directed Cleavage by DNAzymes

Madeleine Meusburger, Martin Hengesbach, and Mark Helm

Abstract

Deoxyribozymes or DNAzymes are small DNA molecules with catalytic activity originating from *in vitro* selection experiments. Variants of the two most popular DNAzymes with RNase activity, the 10–23 DNAzyme and the 8–17 DNAzyme, promote efficient *in vitro* cleavage of the phosphodiester bond in at least 11 out of 16 possible dinucleotide permutations. Judicious choice of the sequences flanking the active core of the DNAzymes permits to direct cleavage activity with high sequence specificity. Here, the harnessing of these features for the analysis of RNA nucleotide modifications by a post-labeling approach is described in detail. DNAzymes are designed such that RNase cleavage is directed precisely to the 5′ end of the nucleotide to be analyzed. Iterative complex formation of DNAzyme and RNA substrate and subsequent cleavage are performed by temperature cycling. The DNAzyme activity liberates the analyte nucleotide on the very 5′-end of an RNA fragment, whose hydroxyl group can be conveniently phosphorylated with ^{32}P. The labeled RNA is digested to mononucleotides, and analyzed by thin layer chromatography.

Key words: DNAzyme, Modified nucleotides, Catalytic DNA, RNA cleavage, Post-labeling, Thin-layer chromatography

1. Introduction

Research on RNA modifications suffers from the lack of efficient methods for their detection in samples obtained from experiments *in vivo* and *in vitro*. As many methods rely on physico-chemical properties, they frequently include a degradation step of some sort, during which sequence information on the detected modification is lost. Since nucleic acid sequences are best decoded by complementary nucleic acid molecules, Zhao and Yu (1)

Ruslan Aphasizhev (ed.), *RNA and DNA Editing: Methods and Protocols*, Methods in Molecular Biology, vol. 718,
DOI 10.1007/978-1-61779-018-8_16, © Springer Science+Business Media, LLC 2011

proposed to combine oligo-mediated recognition, labeling, and physicochemical identification. In that work, RNase H cleavage was directed by a complementary DNA strand flanked by 2'-OCH$_3$ modified ribonucleotides to the 5' end of the nucleotide to be analyzed. The authors point out the possibility of substituting the somewhat cumbersome and expensive combination of proteinaceous RNase and 2'-OCH$_3$ modified oligo by a DNAzyme with RNase activity. DNAzymes or deoxyribozymes are short, single-stranded catalytic DNA sequences. While DNAzymes can catalyze RNA ligation, DNA ligation, oxidative DNA cleavage, and other reactions (2), the first identified and most thoroughly studied reaction catalyzed by several types of DNAzymes is the cleavage of RNA (2, 3). Among these, two types of DNAzymes, namely the 10–23 and the 8–17 DNAzymes and their optimized versions (4–6), have been used both *in vitro* and *in vivo* in numerous applications including modification analysis (7, 8).

Both types of DNAzymes consist of two flanking sequences and a catalytic core region. The flanking sequences can act as substrate-specific arms to bind to their RNA substrate via Watson–Crick pairing. As they can be freely chosen, they allow directing the DNAzyme to the cleavage site of choice. This makes them tools for the RNA biochemist, which are somewhat reminiscent of restriction enzymes in their sequence specificity. However, while sequence specificity of DNAzymes is very high and can be engineered with few limitations, cleavage activity of DNAzymes is much lower than that of restriction enzymes. Thus, low cleavage activity of a given DNAzyme construct indeed constitutes the major limitation in its application to modification analysis.

While the 8–17 DNAzyme tolerates several changes within the catalytic core and still remains catalytically active (6), only few nucleotides within the catalytic loop of the 10–23 DNAzyme can be exchanged without severe effects on catalytic activity, with the boundary region of the catalytic core reacting especially sensitive (9, 10). Both the 8–17 and the 10–23 DNAzyme require the presence of divalent metal cations such as Mg^{2+} or Mn^{2+} for their proper functioning.

For any given DNAzyme, cleavage of the phosphodiesterbond between the adjacent nucleotides N_1-p-N_2 does not proceed equally efficient for all 16 permutations of nucleotides N_1 and N_2. Rather, DNAzymes with different preferences have been evolved by SELEX (6). 10–23 DNAzymes cleave purine–pyrimidine junctions, albeit with different activities. While A–U and G–U combinations are cleaved very well, the activity for G–C and A–C is drastically reduced. The introduction of inosine into the binding domain of the DNAzyme can ameliorate the cutting efficiency of G–C to some extent (5). The 8–17 DNAzymes were originally identified to cleave A–G junctions (4). A few years later, the substrate specificity was extended to N–G junctions (17), and finally to all nucleotide combinations except Y–U dinucleotide junctions (6). Some of the novel 8–17 DNAzyme variants were shown to require Mn^{2+} for optimal catalytic activity. With the recent advances in DNAzyme optimization, the cleavage of the majority of sequence contexts becomes possible. However, the reaction rates of DNAzymes available limit the analysis of nucleotide modifications to the following target dinucleotides: AU, GC, GU, NA, and NG (compare Fig. 1). Thus, by judicious combination of flanking sequences and catalytic DNAzyme motifs, one may sequence specifically liberate a downstream RNA fragment carrying the modified nucleotide on its 5′-end. Since the cleavage reaction produces a 2′–3′cyclic phosphate on the upstream fragment and a convenient free 5′-hydroxyl group on the analyte,

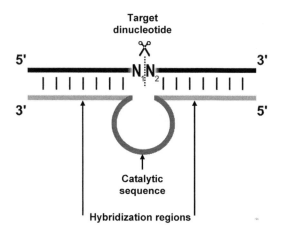

N_1N_2	Motif	Catalytic sequence (5'-3')
AA	8-17, E5112	GTCAGCTGACTCGAA
AG	8-17, E1111	TGTCAGCGACACGAA
AU	10-23	AGGCTAGCTACAACGA
CA	8-17, E5112	GTCAGCTGACTCGAA
CG	8-17, E1111	TGTCAGCGACACGAA
GA	8-17, E5112	GTCAGCTGACTCGAA
GC	10-23	IGGCTAGCTACAACGA
GG	8-17, E1111	TGTCAGCGACACGAA
GU	10-23	AGGCTAGCTACAACGA
UA	8-17, E5112	GTCAGCTGACTCGAA
UG	8-17, E1111	TGTCAGCGACACGAA

Fig. 1. Schematic representation of DNAzyme design. In the *upper part*, the analyte RNA including the to-be-analyzed nucleotide N_2 in a 5′-N_1-p-N_2-3′ context is shown in *black*. The DNAzyme designed to cleave between N_1 and N_2 contains a catalytic core region (*dark gray*) whose catalytic sequence is to be selected according to the nature of N_1 and N_2, as listed in the accompanying table. The flanking sequences (*light gray*) adjacent to the catalytic sequence serve to direct cleavage specificity to the N_1–N_2 site. They are chosen as canonical reverse complement to the respective target sequences.

the latter can be directly phosphorylated with ^{32}P without any intermediate steps. The resulting 5'-^{32}P-labeled fragment is then processed according to the traditional Stanley–Vassilenko sequencing procedures (11), involving gel purification, degradation by nuclease P1 to mononucleotides, and analysis of the latter on TLC. As an obvious prerequisite for this method, the modification must not render the specified cleavage position inert to DNAzyme action. In general, we found modifications that leave the Watson–Crick face of the analyzed nucleotide untouched, such as pseudouridine, m^{5}C or f^{5}U, to be amenable to analysis by this method.

2. Materials

2.1. Nucleic Acids

1. Use Millipore grade water for all nucleic acids and reaction buffers. Store nucleic acids solutions at –20°C or on ice whenever possible at all stages preceding digestion (step 3.4). Touch all samples with gloved hands only.

2. Target RNA: ~5–10 pmol of purified analyte RNA of known sequence, e.g., *in vitro* transcript after modification with recombinant enzyme, or purified native RNA. PAGE purification prior to analysis is strongly recommended. Store at –20°C.

3. DNAzymes of appropriate sequence design, 0.2 μmol synthesis scale, stock solution at 100 μM. Store at –20°C.

2.2. DNAzyme Mediate RNA Cleavage Reaction

1. 5× buffer DZ1: 250 mM Tris–HCl, pH 7.5, 50 mM MgCl$_2$.

2.3. Labeling, Isolation

1. T4 Polynucleotide kinase 10 U/μL.

2. DNase I (RNase-free), 50 U/μL (Fermentas).

3. 10× PNK buffer A: 500 mM Tris–HCl, pH 7.6 at 25°C, 100 mM MgCl$_2$, 50 mM DTT, 1 mM spermidine, 1 mM EDTA.

4. 0.1 mM ATP.

5. [γ-^{32}P]-ATP, 10 μCi/mL. The specific activity is irrelevant, since the labeling will be performed in the presence of excess cold ATP.

6. Urea PAGE loading buffer: 90% formamide, 1× TBE.

7. Elution buffer: 0.5 M NH$_4$OAc. The solution self-buffers near pH 7. Remove particles by passage through a 0.22 μm filter (Steriflip, Millipore).

8. 0.45 μm Nanosep membrane spin filter units (Pall).

9. Siliconized reaction tubes of standard size (1.7 mL).

2.4. Digestion, TLC

1. Nuclease P1 (Sigma Aldrich), 1 mg/mL in water.

2. Carrier bulk tRNA (e.g., yeast or *Escherichia coli* total tRNA, Sigma) at 10 μg/mL.

3. 10× P1 buffer: 200 mM NH_4OAc pH 5.3.

4. TLC solvent A: isobutyric acid:25% aqueous ammonia solution:H_2O (66:1:23 [v/v/v]). Isobutyric acid must not be older than 1 year. The 25% aqueous ammonia solution may be used for 6 months after opening of a fresh bottle. Store in a vented area. All TLC solvents may be prepared in multiple volumes and stored as stock solutions for several months.

5. TLC solvent B: 100 mM phosphate buffer, pH 6.8, and $(NH_4)_2SO_4$:*n*-propanol (100:60:2 [v/w/v]). Dissolve 60 g of $(NH_4)_2SO_4$ in 100 mL of sodium 100 mM phosphate buffer (pH 6.8). Add 2 mL of *n*-propanol and stir until a clear solution is obtained.

6. TLC solvent C: isoprop: HCl: H_2O 70:15:15. Store in a vented area.

7. Fluorescent (e.g., designation "F254") 0.1 mm cellulose thin layer plates on plastic or glass support (Merck or Machery-Nagel). Plates are typically supplied in a 20×20 cm format, which may be divided into four 10×10 cm plates; depending on the nature of the analyzed nucleotide modification, a 20×10 cm format may also be useful.

8. Chromatography glass cuvettes for 20×20 cm plates with a lid.

9. Hand-held UV lamp with ~254 nm emission.

10. X-Ray films or phosphor imager storage plate and readout device (e.g., Typhoon, GE Healthcare) with quantification software, e.g., ImageQuant (GE Healthcare) or Image J (http://rsb.info.nih.gov/nih-image/).

3. Methods

A basic working knowledge of RNA biochemistry and radioactive labeling by ^{32}P is required, including standard techniques such as polyacrylamide gel electrophoresis, and visualization of labeled RNA by X-ray film or phosphorimager.

3.1. DNAzyme Design

1. Based on Fig. 1, identify the target dinucleotide N_1–N_2 in your sequence, where N_2 is the nucleotide to be analyzed.

2. By looking up N_1–N_2 in Fig. 1s table, identify the catalytic DNAzyme core sequence in the last column (see Note 1).

3. Add the flanking sequences to the DNAzyme, which will form canonical reverse complementary double strands with the target RNA. The target dinucleotide does not base-pair with the flanking sequences.

4. Synthesize or custom order the resulting full length DNAzyme sequence, e.g., on a 0.2 μmol synthesis scale (see Note 2).

3.2. DNAzyme Mediated RNA Cleavage Reaction

1. For the single-point analysis of 5 pmol of analyte RNA, compose a typical (see Note 3) 20 μL reaction mixture in a 200 μL PCR tube by mixing.

   (a) 5 μL target RNA (~1 μM).

   (b) 4 μL buffer 5× DZ1 (see Note 4).

   (c) 1–5 μL DNAzyme (10 μM).

   (d) H_2O to final volume of 20 μL.

2. In a standard PCR thermocycler, submit the reaction mixture to the following temperature profile:

   (a) Denaturation at 80°C for 30 s.

   (b) Annealing of DNAzyme and analyte RNA by cooling to 37°C at –0.3°C/s.

   (c) Cleavage at 37°C for 5 min.

   (d) Renewed assembly of active DNAzyme–RNA complexes by repeated cycling for 10–20 times to achieve quantitative cleavage (see Note 5).

3. For optimization purposes (see Note 6), aliquots of the reaction mixture can be analyzed by Urea-PAGE at this time.

3.3. Labeling, DNAzyme Removal, Isolation of Analyte RNA

In this step, the 5′-terminal nucleotide of the 3′-fragment is to be phosphorylated with ^{32}P-phosphate from [γ-^{32}P]-ATP. The reaction mixture from the DNAzyme digestion (Subheading 3.2) can be directly used for this phosphorylation, after which the DNAzyme is removed by incubation with DNase I. All three steps are performed as a one-pot reaction.

1. To the reaction mixture from step 3.2.d, add 1 volume (here: 20 μL) of a labeling mixture (see Note 7) freshly composed of:

   (a) 1 μL (10 μCi) ^{32}P-γ-ATP

   (b) 2 μL 10× PNK buffer A

   (c) 1 μL 10 U/μL T4-PNK/pmol of 5′-ends

   (d) 4 μL ATP (non-radioactive), 0.1 mM

   (e) H_2O to a final volume of 20 μL

2. Incubate the reaction mixture (total volume is now 40 μL) at 37°C for 1 h.

3. For removal of DNA, supplement with 10 U of DNase I/1 μg of DNAzyme.

4. Incubate the reaction mixture for an additional 1 h at 37°C.

5. Meanwhile, cast a standard 7 M Urea-TBE-PAGE of appropriate concentration to isolate the downstream fragment of the analyte RNA containing the 5′-labeled nucleotide to be analyzed (see Note 8).

6. Add 1 volume of loading buffer (here about 41 μL) to the reaction mixture.

7. Heat to 60°C for 4 min, load on gel, and perform electrophoresis until the labeled analyte RNA has migrated approximately 50% of the full length distance.

8. Dismount the gel, expose to X-ray film or phosphor imager storage.

9. Identify the analyte RNA fragment of interest and excise the corresponding band.

10. Place the gel slice in a standard 1.7 mL reaction tube and add 450 μL 0.5 M NH_4OAc.

11. Elute the RNA by moderate shaking at room temperature overnight.

12. Filter supernatant (see Note 9) through a 0.45 μm nanosep spin filter unit into a siliconized 1.7 mL reaction tube.

13. Add 2.5 volumes (~1 mL) of chilled (−80°C) absolute ethanol, mix by inversion, and store at −20°C for ~1 h.

14. Pellet the RNA by centrifugation at 15,000×g at −4°C for 45 min. Remove the supernatant carefully and wash the pellet very carefully once with 200 μL 80% ethanol. Air-dry until no residual washing solution is visible (see Note 10).

3.4. Digestion, TLC

Digestion to mononucleotides and subsequent TLC analysis is a standard method in the field of nucleotide modification.

1. Resuspend the dry pellet from step 3.3 directly in 9 μL 1× P1 buffer, containing 20 μg carrier tRNA.

2. Add 0.3 U of nuclease P1 in 1 μL.

3. Incubate at 50°C for 1–2 h, at 37°C for 3–4 h, or at room temperature overnight.

4. Use a pencil to mark the origin in one corner of a 10×10 cm TLC plate, at a distance of 1.5 cm from the edges.

5. Spot 5,000–10,000 cpm Cerenkov counts in as low a volume as possible onto the origin (see Note 11).

6. Place the TLC in a cuvette filled with solvent A (see Note 12) to a maximum of 1 cm from the bottom. The point of origin is to be situated in the lower left corner.

7. Let the solvent A ascend to the very upper end of the plate. This will take 2–4 h, depending on the manufacturer and the batch of the TLC plates.

8. Remove the TLC plate and air dry it in a well-functioning hood.

9. Turn the plate by 90° counterclockwise such that the point of origin comes to bear in the lower right corner.

10. Repeat the chromatography in solvent B or C.

11. After air-drying, expose the TLC plate for autoradiography and quantification. Use radioactive ink to provide landmarks for later overlay.

12. If necessary (see Note 13), use a UV lamp (254 nm) to visualize and indicate by pencil the positions of the four major nucleotides, whose signals originate from the carrier tRNA and appear as dark spots against a fluorescent background. Use the radioactive ink marks to create an overlay of TLC plate and autoradiography to obtain relative coordinates of the analyzed modification on the reference charts (12).

13. From the autoradiographies, standard quantification software may be used to determine the apparent extent of modification. If known, apply correction factors for the differential cleavage of modified nucleotides and differential labeling efficiency by T4-PNK (see Note 14).

4. Notes

1. Because of potentially differential cleavage activity of modified vs. unmodified nucleotides N_2, the accuracy of quantification is critically dependent on a near-quantitative cleavage of the analyte RNA. From our experience, we deduce that the cleavage efficiency of DNAzymes evolved for the 11 N_1–N_2 combinations shown in Fig. 1 may be sufficient. Because the reported cleavage rates for the remaining five N_1–N_2 combinations are inferior, they are unlikely to be amenable to analysis by this method.

2. If the size of the expected analyte RNA fragment is similar to that of the DNAzyme, it may be useful to increase the size of the DNAzyme by elongating the hybridization regions. In addition, to block phosphorylation of the DNAzyme and to monitor removal by DNase I treatment (see protocol below), it is recommended to order DNAzymes with a fluorescent

dye attached to the 5′-end (i.e., fluorescein). This significantly facilitates identification of the target fragment and allows to directly judge the efficiency of DNase I treatment.

3. When time courses or optimization experiments are conducted, master mixes should be used in all possible cases. Cleavage conditions should be optimized for the target RNA in terms of cleavage temperature, cleavage duration, and denaturing conditions. Generally, the more structured the target is, the higher denaturing temperature needs to be chosen. It may be useful to add a stoichiometric excess (~2-10-fold) of DNAzyme.

4. Certain DNAzyme types may require or work better in the presence of Mn^{2+}.

5. DNAzyme-RNA complexes are thought to assemble in productive, i.e., catalytically active, as well as in inactive conformations. The repeated temperature cycling is intended to permit inactive complexes to dissociate and re-associate as active complexes. We have observed increased cleavage after temperature cycling in some, but not all constellations (unpublished). Note that extended cycling promotes unspecific degradation especially at high temperatures.

6. Useful parameters to optimize upon incomplete cleavage include the concentration of DNAzyme, temperatures and times of the temperature cycle, and magnesium and manganese ion concentration. We have repeatedly observed inefficient cleavage, even of unmodified sites, in native tRNAs. This suggests that the tertiary structure, which is reinforced by nucleotide modifications in tRNA (12), is in competition with the formation of active DNAzyme–tRNA complexes. Therefore measures supporting the stability of the latter complex, such as extending the length of flanking sequences or the use of DNAzymes modified with 2′-OMe or LNA (13, 14) should be considered.

7. Modified nucleosides are frequently phosphorylated less efficiently than their parent standard nucleotide. To reduce biasing the resulting signal during the labeling reaction, a significant excess of ATP over free 5′-OH residues is applied in the reaction mixture, to ascertain maximum phosphorylation efficiency for the modified and unmodified species.

8. Some approximate values are 15% PAGE for a 20mer fragment and 8% for a 80mer fragment. Generally, migration of the analyte RNA fragment in proximity to the Xylene Cyanol dye is desirable.

9. 10–20 µg carrier tRNA may be added to protect against RNases and to aid subsequent precipitation. Because 40–50 µL of the original 450 µL are typically soaked into the gel, 400 µL

are typically recovered. Removal of small gel fragments by filtration (or additional centrifugation and recovery of the supernatant) is critically important, because upon addition of EtOH, the RNA will precipitate on any remaining gel fragments and not redissolve in water.

10. Proper drying of the pellet is important, as residual ethanol may negatively affect dissolution and nuclease activity, while extended drying, especially in combination with vacuum, can cause the RNA to stick to the plastic surface of the reaction tube. Siliconized reaction tubes and the use of carrier tRNA aid in the dissolution for digestion.

11. 5,000–10,000 Cerenkov counts provide a reliable signal for quantification. The spotted volume should be kept below 5 μL, to preferably 1 μL, to avoid damage to the TLC plate surface and accumulation of salt on the spot of origin. A white 10 μL pipette tip allows to repeatedly release sub-microliter volumes by delicately applying pressure to the upper orifice using a fingertip.

12. Although two-dimensional TLC is described here, one-dimensional TLC will often suffice (see e.g., Fig. 2) to separate modified from unmodified species. In a typical setting, the unmodified nucleotide at the target position and the nature of the modification are known. In this case, no

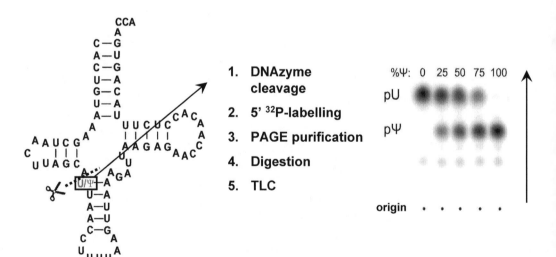

Fig. 2. Example of analysis of Pus1p-mediated pseudouridine formation at position 27 of tRNA[(Lys)] from human mitochondria (16). The tRNA with the analyzed position and the DNAzyme cleavage site is shown on the *left*. The TLC of a calibration experiment conducted with full length tRNA ligated from synthetic fragments is shown on the *right*. Unmodified and fully modified tRNA was mixed in known ratios, as indicated on the *top*.

overlay is necessary because both signals can be unambiguously identified. The appropriate solvent can be determined from literature (15), which supplies numerous reference charts for the three solvent systems.

13. Unknown modifications must be identified from charts with known R_f values (15).

14. Correction factors accounting for differential DNAzyme cleavage and differential 5′-labeling by T4-PNK are determined by mixing unmodified and fully modified samples in known ratios as exemplified in Ref. (8). The calibration factor is defined as the input modification value (in mol modification per mol RNA) divided by the measured modification value. Typically, DNAzyme cleavage and phosphorylation of modified nucleotides will be less efficient than that of standard nucleotides, resulting in correction factors greater than 1.

Acknowledgments

Mark Helm for the DFG funding (HE 3397/4). Martin Hengesbach acknowledges funding by the Landesgraduiertenförderung Baden-Württemberg. We thank Andres Jäschke for generous support.

References

1. Zhao, X. and Y.T. Yu, *Detection and quantitation of RNA base modifications*. RNA, 2004. 10(6): p. 996–1002.

2. Baum, D.A. and S.K. Silverman, *Deoxyribozymes: useful DNA catalysts in vitro and in vivo*. Cell Mol Life Sci, 2008. 65(14): p. 2156–74.

3. Breaker, R.R. and G.F. Joyce, *A DNA enzyme that cleaves RNA*. Chem Biol, 1994. 1(4): p. 223–9.

4. Santoro, S.W. and G.F. Joyce, *A general purpose RNA-cleaving DNA enzyme*. Proc Natl Acad Sci U S A, 1997. 94(9): p. 4262–6.

5. Cairns, M.J., A. King, and L.Q. Sun, *Optimisation of the 10-23 DNAzyme-substrate pairing interactions enhanced RNA cleavage activity at purine-cytosine target sites*. Nucleic Acids Res, 2003. 31(11): p. 2883–9.

6. Cruz, R.P., J.B. Withers, and Y. Li, *Dinucleotide junction cleavage versatility of 8-17 deoxyribozyme*. Chem Biol, 2004. 11(1): p. 57–67.

7. Buchhaupt, M., C. Peifer, and K.D. Entian, *Analysis of 2′-O-methylated nucleosides and pseudouridines in ribosomal RNAs using DNAzymes*. Anal Biochem, 2007. 361(1): p. 102–8.

8. Hengesbach, M., et al., *Use of DNAzymes for site-specific analysis of ribonucleotide modifications*. RNA, 2008. 14(1): p. 180–7.

9. Zaborowska, Z., et al., *Sequence requirements in the catalytic core of the "10-23" DNA enzyme*. J Biol Chem, 2002. 277(43): p. 40617–22.

10. Zaborowska, Z., et al., *Deletion analysis in the catalytic region of the 10–23 DNA enzyme*. FEBS Lett, 2005. 579(2): p. 554–8.

11. Stanley, J. and S. Vassilenko, *A different approach to RNA sequencing*. Nature, 1978. 274: p. 87–9.

12. Helm, M., *Post-transcriptional nucleotide modification and alternative folding of RNA*. Nucleic Acids Res, 2006. 34(2): p. 721–33.

13. Schubert, S., et al., *RNA cleaving '10-23' DNAzymes with enhanced stability and activity*. Nucleic Acids Res, 2003. 31(20): p. 5982–92.

14. Vester, B., et al., *LNAzymes: incorporation of LNA-type monomers into DNAzymes markedly increases RNA cleavage.* J Am Chem Soc, 2002. 124(46): p. 13682–3.

15. Grosjean, H., G. Keith, and L. Droogmans, *Detection and quantification of modified nucleotides in RNA using thin-layer chromatography.* Methods Mol Biol, 2004. 265: p. 357–91.

16. Voigts-Hoffmann, F., et al., *A methyl group controls conformational equilibrium in human mitochondrial tRNA(Lys).* J Am Chem Soc, 2007. 129(44): p. 13382–3.

17. Li, J., W. Zheng, A. H. Kwon and Y. Lu. *In vitro selection and characterization of a highly efficient Zn(II)-dependent RNA-cleaving deoxyribozyme.* Nucleic Acids Res., 2000. 28(2):481–8.

INDEX

Ruslan Aphasizhev (ed.), *RNA and DNA Editing: Methods and Protocols*, Methods in Molecular Biology, vol. 718,
DOI 10.1007/978-1-61779-018-8, © Springer Science+Business Media, LLC 2011